MAN IN ALL THAT IS:
On How the Universe's Order Enters Our World.

To Rose Ann —

Very Best Wishes.

Jim D. Alexh

Thank you for your
Kind Hospitality at
The Morrill House.

4/5/87.

MAN IN ALL THAT IS:
On How the Universe's Order Enters Our World.

by
Ivan D. Alexander

Peter E. Randall
PUBLISHER

Peter E. Randall Publisher
Box 4726, Portsmouth, NH 03801

ISBN 0-914339-10-9

Habeas Mentem: To Have the Mind (Latin)

There will never be such a thing as a writ of habeas mentem; for no sheriff or jailer can bring an illegally imprisoned mind into court, and no person whose mind had been made captive . . . would be in a position to complain of his captivity.

Aldous Huxley,
Brave New World Revisited

CONTENTS

Book I: To Have The Mind

1. Introduction: Is There a Natural Order?..........1
2. Let Us Create an Idea.......................................8
3. What is the Form of Interrelationship?..........15
4. Each One of Us...22
5. Each One of Us, Conscious............................29
6. We Can Choose...37
7. Having a Mind, We Have a Right.................45
8. A Person in Agreement.................................54
9. How Do We Build a Society?......................62
10. A Society Built on the Principle of
      Habeas Mentem..69
11. A Society of Individuals..............................77
12. In a Mechanism of Exchange......................85
13. How Do We Measure Value?.......................94
14. Wealth is a Conscious Act..........................103
15. In a Society Conscious of Itself..................112
16. Epilogue: The Mind is the Communicator...122

Book II: To Have The Soul

17. How Do We Know — "Who Am I?"............133
18. What Is the Energy? On Belief.....................140
19. Let Us Remember: Images From
        the Right Side of the Brain.......................148
20. The Idea: All That Is.....................................156
21. Patterns of Triads: Being, Energy, Idea.........163
22. When We Are in the Mind:
        We Create Reality.....................................170
23. We Are Conscious: We Give.........................177
24. Within Our Reach: Habeas Mentem II........185
25. The Given Word: Who We Are....................193
26. Let Us Work Together: On Friendship
        and Love..200
27. It Is Important to Have the Soul.................208
28. It Is Important to Have Faith......................215
29. It Is Important to Love One Another...........221

. . . grasping the stuff that dreams are made of *and shaping them into logic, into rules of life and conduct . . .*

— Juwain, Martian Philosopher

Book I:
HABEAS MENTEM, To Have the Mind:
A Metaphysical Reality

Chapter One

Is There a Natural Order?

IS THERE A NATURAL ORDER in the world of man, or is all that appears organized around us the work of the mind of man? Is society a natural manifestation, or is it an artifice man has created for himself for his own purposes? All the knowledge man has amassed and incorporated into his present civilization, the philosophies and sciences, the institutions that promote this knowledge and expand it, the mechanization that harnesses our present understanding of the reality around us for our livelihood and comfort; are they all human creations exclusive of some Divine Order in a universe that has accidentally become what it is? Or is our universe a natural order moved by its own knowledge for its own purpose? Who is man?

To observe the world, to become a part of it and live in its environment, to suffer the pain that drives the will to observe itself in reality, to observe the reality that immediately affects us is to think. We are moved to ask questions and explore the world we live in. Are we moved from within or without? Is an inquiry the mind's response to a set of circumstances it happens to find itself in, responding to how the body's mechanism is affected by it, or is an inquiry an aperture through which the mind and the universe communicate? For us to ask the question "Who is man?" requires that there exist a stable point of reference against which the identity of man can be defined. The question presupposes an order out of which can be constructed the answer. To answer "Who is man?" from a void presupposes a universe that is meaningless and one which cannot reply to the question. If we presuppose a meaningless universe, then the question is pointless and we must answer merely from our observation of ourselves that man is such and such. But to live is to think, and when the observations have been made, the question remains. We persist with our question of our identity: Who is Man? We live and thus we must ask.

We have made great strides in describing the physical universe we live in. Our astronomers have successively penetrated deeper and deeper into space with their understanding of the cosmos. Our physicists and chemists have explained with confidence the nature of physical matter and the laws that govern its behavior. Mathematicians have created languages that describe physical reality. Engineers have harnessed much of our understanding and have developed a technology which to an individual can appear awesome but which can be employed to serve us. Bio-scientists have successfully combined their efforts with the medical profession to enable us to prolong our body's life span. We have become sophisticated in our physical knowledge and can feel confident in our beliefs. Speculations are tested and discarded if they fail to yield the expected results. A truth is unacceptable until proven. We have abandoned our ancestors' gods for our new knowledge of the universe, which through tests and proofs has yielded an image of the universe as a great, precise machine. We have established a great foundation of physical truths, some of which we have applied to ourselves successfully. We have mechanized ourselves and have tried to occupy that image. To some extent, this image of a modern, mechanized, technologically sophisticated man fits. We can live comfortably, if at times awkwardly, in an environment that somewhat resembles the inside of a space craft insulating the soft mechanisms that is our physical body from the harmful forces of our cosmos. In this sense, we are no longer primitive; we have the power to create our own environment. We can create order.

If this new found power has led us to conceit, we can be forgiven. We do not know that our accomplishments are not in themselves the greatest achievements of the mind. We have no reason to believe that we have not improved on the universe and its primitive chaos. If the universe is an accidental collection of forces that have been manifested into what we observe today, then what man has wrought from this accident is truly a great accomplishment. If science and the mind of man can create order from disorder, then they are great and understandably can be deified. The mind of man then is the light that relieves the darkness and is in itself its own salvation. To turn to religion and other worldliness, to seek for a meaning greater than that of the present existence, is to become sentimental, tolerated as one would tolerate a child's fond imaginings. We can have a fondness for a belief or an idea that can bring us comfort against our fears of the unknown and ultimately of death. Religion can be tolerated to fill those needs, but it can have no place in the hard decisions with which we must run our world. The mind of man is the supreme manifestation of order in our reality and its intellect is the greatest power in the universe. It is a logical conclusion from the successes of a logical interpretation of our physical reality. To succumb to a belief in order greater than that created by the intellect of man is weakness and should not be tolerated when hard decision must be made. To believe otherwise is to be sentimental and weak or, worse, to be ignorant and primitive. The mind of man succeeds against the universe because it is in and of itself autonomous and supreme in its intellectual accomplishments. Is it?

Even if Earth were not alone as an inhabited planet and there were civilizations throughout the cosmos far superior in their accomplishments to what we have here, the maxim would hold. The mind is supreme and in time will achieve what others have achieved before us. If it were found that more advanced civilizations regarded a Deity, we would consider it a luxury, for which we can still have no room on our planet plagued by so many pressing problems. In time, we too could be tolerant of such an indulgence and contemplate a Deity, but we have no leisure for such now. The world is plagued by problems and we have no time to divert our efforts from our responsibility to it. The world is the intellectual burden of the most intelligent, the gifted, the best educated and the most fortunate. The system that runs the planet must be tightened and better controlled. Human activity must be better regulated to yield a greater product. The wants and needs of the people are increasing and must be met if we are to avoid a global disaster. The problems are pressing and the mind must suffer no diversion. It sees itself destined to succeed over chaos and avert catastrophe. But why "destined?" Why not let catastrophe overcome and return the elements to their origin of a mechanical if disorderly, accidental universe? Why must the mind work so hard against what appear to be overwhelming odds? Who is man?

If we are denied the salvation of the Spirit in our pursuit of the hard reality, then we must turn to an identity that is consistent with the way we perceive our universe. If we see the universe as soulless, the breath of life but a phenomenon of probability that has materialized itself in the life forms of our planet, then what had hitherto been ascribed as characteristics of a soul must be the creation of our imagination. The qualities of the soul must be what the mind had created for itself to adorn what would otherwise be its naked existence. We have created the soul in our image to appease a cold universe. What life and reality seemingly cannot bring, a feeling of oneness with everything, a belonging to an order greater than ourselves, we have created in the image of the Spirit. We have tried to unite our life into the matrix of a greater Self, one that is eternal and can transcend that brief moment of cosmic time that represents an individual's life. But our modern perception of reality does not allow us to pursue this oneness seriously. We can yearn for it, but we cannot truly believe it. Our innermost thoughts may secretly speculate on a greater existence, but our disciplined thoughts return us to the material world. Our greatest achievements are man made, they are not miracles. Our salvation is a product of our thought and labor, not divine intervention. We are inescapably drawn to the conclusion that the mind is supreme, not the cosmos. The identity of man is defined by the reality of his thoughts, of "I am", the reality of his consciousness and not by a surrender to a being beyond. "I think," "I am," and thus "I am the consciousness that is my identity." If the conscious mind is colored by altered states or by substrata of the conscious and subconscious, then they are only shades of the oneness that composes the identity of our being. In the end, we have defined ourselves in ourselves and have closed the door on speculations of our identity outside our being. The mind defines our identity and the soul is entirely locked within this new, aware, sophisticated modern mind. It creates; it controls; it is progressive and thus it is confident in its position in reality. It is itself, an autonomous, cognizant and organized entity in a disorderly, probababilistic universe. The mind is a closed system, a sealed bubble within a chaotic and potentially hostile universe. It observes, it controls, but it is nevertheless autonomous.

When man first caught glimpses of "I am," a new phenomenon occurred. The unconscious or semi-conscious state of the animal suddenly became aware of itself. The evolution of the mind had just passed a major threshold of its future development. Man could now think of himself in the abstract: "I am." The animal mind became human; it began to think. If we are to speculate that it thought of itself first, then we must allow for the same mental mechanism to think of the world around it. In its primitive way, it began to philosophize. What its first conclusions were we can only infer from records of our history or from observing the remaining primitive people of our planet. The world became inhabited by spirits, magical forces that have the characteristics of personalities and the power to alter events. These spirits became man's definition of what his mind perceived as reality, and this perception persisted even into modern times.

In the civilized world, the philosophy of spiritualism is but vestigial, but it persists nevertheless, if in an altered form, in our modern religions. But man still sees himself only as "I am." The evolution of his philosophy has progressed his thought beyond imposing his traits of personality on all aspects of reality, but not beyond. Man has philosophized himself out of spiritual context in the universe, his mind abstracted from the events of the cosmos, but he has not progressed beyond the risk that is the fleeting moment of his life in reality. That we live and understand we live is still a miracle. We no longer see the world through the heathen's eyes, but nor do we see the universe beyond its apparent chaos. Events are perceived randomly, mechanical but of no known consequence. The throw of a die or the assumption of a venture can be calculated as to the risk involved, but no certain outcome is evident. Though we have progressed and learned to harness many secrets of our physical universe, we have not progressed beyond "I am." We have penetrated deeper into the outer and inner dimensions of space but have nevertheless remained autonomous in our self perception. We have evolved beyond ascribing the traits of our personality to reality but have not evolved to the level that defines our personality in terms of the reality within which it is evolving. We still cannot define our identity. "I am" is still a self-enclosed, incomplete realization divorced from the world outside itself. It is a recognition of the self in a universe that in many ways appears strange and alien to us. But we came from there. In it is our beginning and our development. Why is it alien?

The human mind is unique in that it can speculate about itself. It can speculate about itself in relation to everything around it. The creative mind can imagine relationships that may or may not exist in reality and revert these relationships back to its origin, the self. Our sophisticated imagination allows us the versatility to philosophize the relationships away from the self and lead us into the conclusion that "I am" can be itself unrelated to reality and accept this new premise. We can progress to the point where we can even deny our existence. The freedom with which we can pursue these thoughts is the beauty of the versatility with which our mind can function. But "I am not" violates our experience in reality and can be accepted with even less credence than the spiritual world of our ancestors. We are not disassociated from reality, though we may at times wish to be. "I am not" is not our identity, though it can conceivably be a state of being in our universe, if not at the limit, at least approachably so. But the mind that can create itself its own non-being should have equal freedom to theorize itself in being. Awake, aware of itself, its identity could be discovered by its wide roving imagination. We came from there, and somehow, out there, we became the Human Being. Who is Man?

We, our minds, our civilization are at crossroads. We can choose to remain in "I am." It has served us well; it brought us from our primitive self to our present successes. Or, we can go beyond it to seek our identity and answer our question "Who is Man?" We can seek within ourselves deeper still for clues to who we are, or we can reach out into the cosmos for a relationship that is our identity. We have looked into ourselves and become man; "I am" is our human element. But it is not our human identity. If there is a natural order, then we are more. We must find what that more in ourselves is. To look into ourselves without finding ourselves out there, in reality, can be counterproductive and lead us into a sterile identity divorced from reality. Divorced from reality, we can then quickly regress into a negative identity which, brought to its extreme, would mean our extinction. That route threatens us as a specie, though it is unlikely that the specie would follow it to a man. "I am not" may be entertained casually, but it cannot be pursued seriously. The survival of our planet's human identity is too serious a task to be approached casually. Our development requires hard answers to pressing questions if we are to avoid future social chaos and self mutilation. Our world is faced with many pressing needs which cannot be met by the successes of our past accomplishments. Reality is quickly closing in on our mind's lack of self direction and is forcing us to find ourselves. We must find ourselves or reality may reject us; our civilization may perish into history. Such is the risk that confronts us to progress beyond "I am." "I am" was our first identity, our first awakening. We must now probe reality with our imagination, beyond our present accomplishments, and find "I am Who?"

Chapter Two

Let Us Create an Idea

LET US CREATE AN IDEA that can think itself. It is a simple idea, of childlike simplicity; yet, it is complex enough to describe the mechanisms of a universe. The idea is that everything starts with "three." It is that in the beginning space was divided originally by but three points of existence. These three points were the only thing in existence in this primordial space and nothing else could have existed besides them. It is only an exercise in a creative idea, but these three points could be thought of in no other terms, not compared to anything, except their own existence. Hence, in a universal void, these three points existed autonomously, at some undetermined distance from one another, and entirely alone. They had no mass and no defined relationship with regard to one another. The universe was then in its utmost infancy and was about to give birth to a new idea: a relationship of three points somehow interrelated to one another. Hence, an interrelationship. From these basic three points, this basic triad, now formed an idea that could think itself in terms of itself. It is the fundamental idea underlying the rest of this text, the universe's first interrelationship.

Let us now start with a simple everyday observation: that nothing ever exists alone, in a vacuum related to nothing else. No matter how small and insignificant or how great, it is but a part of a matrix that imposes itself on its existence. Even brought to an immediate present, the chair I sit on, the pen in my hand, the room, the trees in the woods, the distant stars, the planet, the past . . . they all are interrelated to one another somehow, simultaneously, infinitely and minutely. These relationships may be sometimes entirely meaningless to us, inexplicable in rational terms. I do not know that if I step on an ant I am not sinking a ship at sea, though it is unlikely that the universe works that way. However, there is some cosmic connection between all things. Thus, though we may be unconscious of it, all things are always somehow related to one another in this vast matrix of things we call existence. This observation may at first sight seem minor in importance, but we will see shortly that this idea of interrelationship is a powerful concept capable of spanning the understanding of being in the universe.

Thus, let us start with our idea of three points in space and make them the basic idea. All interrelationship starts with three. Two things can be related to one another but a third is needed for an interrelation. If we illustrate this as three points in space, then we can immediately see that they must be somehow, randomly positioned in space. They occupy some position in relation to one another, though it is impossible to gauge how far they are from one another since, at this state of development, there exists nothing else with which to measure distance. We can know, however, that the three points do have an internal relationship, an interrelationship that results in the form of a triangle. By definition, three points form a triangle. This is always so, anywhere and at any distance in the universe. If space is curved, then the resulting triangle will be equally curved but it will not negate the definition: three points in space are interrelated into the form of a triangle. Thus, this triangle is the first interrelationship of space.

To examine this idea further, we can also observe that each side of this triangle forms a line. Projected into infinity, either end of this line dissects the universe in two. If the line is projected into space as a plane, the universe is further cut into two halves. But these are mental digressions. The main point of interrelationship is that they are the only things in space that are working on themselves and on the space around them. Because nothing else exists, all definitions within this space must be in terms of our three points. Firstly, each point has only the definition of itself in terms of itself: a point. Secondly, because of interrelationship, each point also has a definition that is relative to its position in terms of the other two points, in terms of their interrelationship. As such they can no longer be defined merely as points but must be defined as being within the resulting interrelationship of the three points. They are a point within a triangle. They are somehow related to one another. They are also related to the resulting totality of their interrelationship, their space, their relative positions. Thus, each point has but a minor meaning in and of itself, whereas it has a greater meaning in terms of the other two points and their interrelationship. Finally, because at this stage nothing else exists, each point is defined and can be understood only in terms of the other two points and their resulting shapes and patterns.

We now have a new way of looking at things: each thing in existence, whether it be a single point or an atom or pebble or planet or galaxy; each is defined by the interrelationship of everything else. Space, the universe, is filled with an infinity of interrelationships as here illustrated by three points. Each thing is somehow interrelated into everything else and from this interrelationship flows its definition in terms of that interrelationship. In our initial illustration nothing existed but our three points, thus each point could be defined only in terms of the other two points. If there had been instead an initial number of a hundred points, then each point would be instead defined in relation to how it was in the interrelationship of the other ninety-nine points and their resulting patterns. However, in the real world, the interrelationships are innumerable, and yet the idea still holds. The more complex the interrelationship, the greater its dimensions, the greater the definition of each thing within it in terms of the whole. Each thing within this whole becomes defined in terms of everything else around it. This is merely a function of an interrelated arrangement of things. Thus, in the real world where the dimensions of these arrangements grow to infinity, each thing becomes defined by how it is interrelated to everything else, ad infinitum. In effect, the being of everything else to infinity defines it in relation to this infinity. Each thing is exactly as the state of everything else allows it to be. At infinity, the interrelationship definition of our universe is brought to completion: Everything is Where and What it Is because of Where and What is Everything Else. This is without exception, for all is now part of an interrelated whole. The force of interrelations has rearranged itself exactly in relation to how has been everything else from the beginning. Now, it is itself. No one stirred it from without. The composition of this universe is entirely self contained, by definition, for otherwise it would not be the ultimate totality. From this vast infinity of universal interrelationships now flows a new definition of an idea that can think itself: It is defining for us a single point in space.

Each thing, at every moment of time, is interrelated to everything in infinity and back. At infinity, the arrangements of the universe that have moved themselves into position have moved a thing into the position it is in the present. But that position is meaningful in a way greater than merely its place in space and time. If all things are assembled to infinity, then from that vast assembly is a definition for each thing in reality. It is a definition that is communicated from infinity in terms of how it is positioned within it in terms of the universal totality. It is an awesome thought, for if nothing exists outside this totality then how it is in reality is how it is defined by the image of that totality. If no motion exists outside the totality, then how anything within reality had moved to its current position is how it was placed in terms of everything else. Thus, each thing is moved within the universe, if it is not self propelled or moved by an outside influence, as it is within the totality image of the universe.

We still do not know what this means, for we cannot fathom a totality image of infinite interrelationship. But the idea works on. It is a simple system that can span infinity, our reality. Though our understanding of its complexity may fail us, we can know with confidence that this same complexity is working away from us in all directions of existence, beyond where our mind can currently understand. The greater the dimensions, the greater the complexity of the arrangements and the greater the definitions on any single one thing. We have come a long way from our primordial three points. However, if it is difficult to understand interrelationship on a fairly small scale, i.e., the natural forces of a planet, think how much more complex are the interrelationships of a universe. But the basic mechanism of our triad works on. Independent of how we think of it, interrelationship works on.

Interrelationship spans all infinity at every moment of time, instantly. Any change in reality is instantly recorded in the definition of what it is from totality and is instantly rearranged in its new meaning. The change in interrelationship defining the new arrangement instantly communicates to infinity and back. This communication with reality is instant but it simultaneously spans time. The relations that were our universe moments ago are in the process of becoming the relationships that will be reality moments from now. But it is also that time is interrelated into the whole. If the universe is self contained, then it follows that interrelationships can be understood either along their space or time axes. In an instant can be grasped the entire interrelationship of the universe; in the totality can be glimpsed each change of time. In interrelationship, time and space are but facets of one another. They are related to the whole whose conditions of existence are determined at infinity. If there are no outside movers, then past and present and future are but conditions predetermined by the arrangement of everything to infinity. This would leave scant room for free will and would suggest that the future can be known. However, as we will see, at infinity, this condition has room to be altered by how we perceive reality. We still do not know what this means at infinity, what this idea looks like in its totality, but we have an idea that can begin to define for us what the space-time relationship of infinity looks like in our reality.

Thus, let us think of interrelationship as not merely an idea but as a real force that exists between things, that defines their being and is responsible for the flow of events. This force is defined by the conditions that have been presupposed by the arrangement of everything else beyond the thing or event defined. Every natural arrangement or circumstance is exactly the way it is because that is how the universe's pressure of everything else at that moment allowed it to be. Being is a conditional thing: It is conditional upon the state of being of everything else. Now, think of this state of being as infinitely complex and of infinite dimensions. Nothing can exist beyond this complexity, nothing is random within its totality, and nothing can possibly escape from this infinitely complex inter-relationship. It is the state of being of reality. Its state of being "knowledge" of itself is complete, knowing itself infintely and completely. It "moves" from within in terms of everything else. Its motion is never random but always coordinated exactly as defined by all the other conditions of existence within it. The "Whole" determines each one of its infinite parts, and infinity begins to take on a new dimension, a kind of super being of interrelationships defining itself not only in the most minute detail but also to the edges of its greatest, most inter-galactic potential. Infinity is an interrelationship that takes on a new characteristic: It is the reason, the responsibility for the way it is within itself. At the limit, at the totality that is infinity, this responsibility becomes more than the itself, it begins to grow into the future. If we were to equate an interrelationship to be-ing similar to a thought, then what we are here describing is a universe that is moved by its own "idea" of itself. It is a universe that can "think" itself, infinitely so. All possible mathematical formulas, all laws of universal physics, all the known knowledge of the mind of man, all probabilities, are "known" in the way the interrelated super-structure of reality that is moving itself is in a way determined by its greatest totality. It is an "Idea" aware of itself in more than merely the level of ideas, for it is the force that defines it physically. Infinite inter-relationship is the physical idea that moves itself. Think of a universe thinking itself. These are the dimensions that inter-relationship is able to span. The result is physical being.

Though we can but begin to conceive of such an idea, we still cannot know what an idea that can think itself is thinking. We can think of it theoretically, think of its structure, but we cannot actually envision an idea actually thinking of itself. We are limited by our three dimensional world to seeing things as they are presented to us in our immediate environment. We cannot, for example, envision the by-axial interrelationship of space and time simultaneously. But the universe's mechanism can and does, and in doing so defines for us our everyday reali-ty. We can perceive reality only in a very small portion of the universe's totality; the universe can see itself as a total reality in its entirety. To this entirety it responds and moves itself in that entirety's image. The power to do so is thus given over to the "mind" of the universe. We are simply observers, our minds struggling to grasp the concepts that are defining this reality. But if our thoughts can project themselves only into this dimension of reality, our thought's creation of an idea of inter-relationship is free to transcend reality into the realm of a totality, multi-dimensional universe. We have given the power to our thoughts to propel themselves beyond our limited dimensions and understanding off into the vast reaches of reality beyond our conceptual ability. Thus we have started something: Same as we cannot reach conceptually the limits of infinity and our minds cannot interpret definitions from infinity that our new thoughts are creating for us, through the forces and cir-cumstances that are displaced by the atoms of our being, we displace that portion of reality that is our presence, our human state of being. We still cannot know what that means, what the universe is "thinking" of it, but we can speculate on what it is we have created ourselves. This is the thesis underlying this book. We will explore free will as well as our relative space-time interrelationship definitions. We will project ourselves in-to space and come back with new definitions of ourselves as a living, thinking, and conscious specie. By thinking of these things we will have launched ourselves into those dimensions we are unable to span conceptually with our minds. We still do not know what this means, but in some small way we have intruded our being into that state of being that defines all things. From our basic triad, from our basic three points in an empty space, we have launched ourselves into that infinite reality we are striving to span. Imagine an idea that can think itself!

Chapter Three

What is the Form of Interrelationship?

WHAT IS THE FORM of interrelationship? What does it look like? What does interrelationship mean in our real world experience? How does it affect reality?

We can now see our world in a new way. We need no longer view everything in it only symbolically. The abstract names we have given things in our world can now go beyond symbolism meaningful only to us. We have seen things simply, childlike putting together bits of knowledge into a puzzle. From the general knowledge passed down to us, we have given meaning to things in the world by agreeing on a thought, a word, that would describe the thing to us. From these we have assembled these words, concepts, into a complex of ideas which help us to understand what everything is. We have done this naturally and successfully; it is our everyday experience. Thus we have built up an elaborate network of ideas which results both in a practical understanding of reality and in philosophies which stem from this understanding. In our building block fashion we have proceeded from the small to the large, the simple to the complex, yet always maintaining mastery over our symbols. Our definitions of reality have been our creations, creations we would not release from our control, thus making us masters of our world. To now, understanding rested ultimately in its acceptance by a consensus, not on how ideas existed and worked in and of themselves independently of us, but as accepted by a general agreement testing the results of these ideas. We thus perpetuated a uniformity in the way we saw things. By doing this, from the dawn of our time, we have "named" things, given them a definition which had placed them in a certain perspective within the universe. Things were not defined intrinsically, to our knowledge, in and of themselves, but defined by the symbols and names we had chosen them. Thus, to our minds, the universe did not exist as a definition of itself, in its own right, but rather as a definition from us; we were the creators of reason within it. We named it and from our bits of knowledge assembled a coherent, philosophical whole. But now, we can see things in a different way.

Things can now have a meaning of their own. They can now fit into a complex of understanding that is intrinsic to them, if not entirely to us. The definition of each thing in reality can now be determined by reality, in our thinking, rather than a meaning formed by our general consensus. We can and have been known to be wrong in our thinking about reality; reality can now have the right to be free from our error; it is always the ultimate judge. Rather than us being the creators of meaning, we can now see that in our creations, we are nevertheless only observers of the meaning reality is giving itself. In reality, each thing is defined by its presence in the complex of everything else. We can now see it this way, through interrelationship, and see what reality is saying about itself.

Things exist through association. The pebble lying on the beach is smoothed and rounded by its environment. It is tossed with each wave, chipped minutely against other pebbles, polished by wind and sand into its present smoothness. Its shape is exactly as it is because that is how it has been molded, through time, by the presence of all the circumstances that have been and are its environment. If it is nearly round, then it could have been made flat or oval; but it was not. From the distant reaches of space and time in reality came the forces and circumstances that, with each passing wave, had lifted it and caused it to be struck again, each blow molding it minutely, from all directions of reality, into its present, exact shape. It is there as all the interrelationship of forces have made it, then. Now, if it is at the water's edge, its origin may have been imbedded in some distant mountain; if it is textured smooth, it may have been jagged and coarse; if it is light or dark, flawed or perfect, it is exactly as it is. On the beach, it is exactly where it lay. No hand reached down to lift it and toss it further. It is itself, there, embodying in its form the full history of itself from its beginning to now. Not a moment was lost nor an experience not etched on its surface. It is what and where it is from the beginning of its time to now, perfectly, in terms of everything else.

As we lift our observation from the pebble and its interrelationship to the beach, the waves, the ocean, shore, the planet, its solar system, galaxies, and off into the dimensions of our universe, we can see that pebble in the image of cosmic interrelationships, out there, that defines it as it is. That state of being, made more pronounced and stronger with the growth of its dimensions, with the interconnectedness of all things in reality, has sculpted delicately the smoothness that is our singular pebble. Somewhere, in that vast, galactic complex of interrelationship, is the image that defines that small, round pebble. At that moment, looking back at infinity from the exact position of that pebble on the beach, untouched, it is itself in the image of infinity as that image has allowed it to be. That infinite pressure of forces and circumstances that has molded it through time has done so exactly, infinitesimally from the reaches of infinity, and molded it into its present image. What does it look like at infinity? Untouched, it looks exactly like itself.

The pebble is a definition of space and time at that point of infinity. What does infinity look like there? It looks like a pebble, then. It is a mirror image of itself there, then, solidified in our physical reality. That point in space and time has other definitions but, as a physical definition, at the limit, it is a materialization of a pebble. Where, at what dimension of infinity, is that image formed? Is the totality of that definition small or is its interrelationship approaching infinity? How powerful is the composition that is stone?

The compositions of atoms, molecules, and crystals, together as stone are very old. Their image is an ancient composition in the matrix of reality. The more stable an interrelationship at infinity, the more durable is its definition as an object of reality. The pebble is a hard, stable substance composed of atoms and crystals whose transitory nature is only evident in the changes of the stone's shape. Then, the totality of the interrelationships that defines its substance must be greater and older than those that define its shape. It is an old universe that has had the time to stabilize the matter of its basic building blocks. Its changing shapes are newer images of interrelationship. Yet, the space that is between the atoms that have been organized into that stone, the values of interrelationship that define that space, is the same space that exists outside the stone in all directions of space. The space-time dimensions of the totalities that progress from the stone's basic substance to the definition of the pebble's final form is only a progression of degree. One is older, the other newer; one is perhaps greater, the other still less stable, formative; but they are all facets of the same principle: the universe's definition of itself. Intricately moving vast forces within itself, the universe can define in the minutest detail each one of its interior substances in the form of its greater images. What is inside the stone, its crystals, its molecules, is only an interrelationship that predates the form that is on its outside. Together, they form the definition of a pebble that, if it could look out into infinity from where it lay, in the vast images of the cosmos, it would see itself.

If we could enter the pebble and shut ourselves off from the world outside it and call its outer form and dimensions infinity, then we would be in a universe within itself. Each atom would be in relationship to every other atom, interrelated to all the atoms within its total form, each contributing to its definition as a whole and each thus defined by that whole. Its gyrations within the whole would be modified by the substance that is arranged around it. The pebble's shape will exert a certain influence on the forces that affect each of the atoms in relation to the others. Because the substance is a solid, it may seem unlikely that this influence is meaningful, but if we enter the pebble deeper, into where this solid is now seen as but an interplay of forces in space, and deeper still where the space between the atoms takes on astronomic dimensions, then the forces that are created within this space respond to the outer shape, outer dimensions, of the pebble. Negating outside forces, as if the pebble were the only body in all space, then the outside dimensions of that body would become the totality image of each of its parts. Each atom would then be in some way in terrelated to every other atom in an image that is the pebble; each atom would be acted upon by the image of that totality. If the pebble were moving in space, then the atoms and molecules would also be gyrating and vibrating in the direction of that space.

Perhaps, the pebble being but a very small segment of reality as we know it, this influence would be minimal, though we do not know that each molecule within this pebble does not carry in itself an image of its totality, much as a fragment of a hologram carries in itself an image of the whole picture. But there is a universe outside to which the pebble must defer. The atoms and molecules within the pebble, though they are protected by its outer form, are nevertheless acted upon from the greater space outside. If we could see infinity as it would look from inside that pebble at the point that is an atom, we would look past the pebble and see the image that is the atom iself. Thus, at that point in space, the definition of infinity is an atom. Such is infinity; through interrelationship, every point in reality is exactly as it is in its greater image, at infinity. Reality is the definition of infinity as seen there.

Then all things are at infinity as they are. It is a property of our universe that each thing within it is inscribed somehow in its vast network. The atom is not an isolated, detached singularity of our reality; it is inherent in its total meaning. At the outer dimensions of our universe is the definition of the atom focused at each point where each atom exists. It is a property of our universe that what we see in our physical reality is what the arrangement of everything else at the outer limits of our universe means. Those properties exist out there as a vast image of itself. Nothing can exist that is not inherent in the meaning of the universe.

What we see is its real image. It is not some imperfect shadow of reality; it is reality. The question arises: Do we see it exactly as it is? We still do not know. The subjective mind strives to influence reality with its own design. But we can know with confidence that whether or not what we see in our mind through our senses is the exact image, the object we are observing is the reality defined there by interrelationship from the infinite reaches of our universe. At each point of reality, the image of infinity materialized there is the object we see. In a world untouched by hands, this is always so. Each thing is defined physically, at its point in space and time, as that point is defined at infinity. We can give it a name, but now we are only the observers. Reality already "named" it; our name is only a shadow of that name. As we name it, reality is naming it physically, molding it into its present form. If we do not touch it, then we do not affect its real definition from infinity. There is a real definition of things. It is defined by its meaning in the universe materialized exactly as we see it.

If we do not touch it, unobserved, the meaning is actually more itself. If we could turn our mind away from it, not influence it, then it will not be moved in its definition within its matrix; its definition will be perfect. The forces and circumstances that have affected it from the beginning of time will have worked it exactly, undisturbed, infinitessimally into its present image. Think of it; at infinity it is almost a memory of itself. Undisturbed, the meaning inscribed in a natural form dates back to its beginning and its experiences registered in all dimensions of space. Untouched it is itself perfectly in its natural, almost supernatural, setting. Void of any outside influence, the universe is a natural setting of its own image as it moved itself and thus materialized in the physical reality. As a part of this reality, we are then able to observe it.

The universe is its own sculptor, but once sculpted, each form within the matrix of reality has a special meaning in itself. If we could see it with our senses and mind the same way it is seen by the universe, its form would take on a special meaning, almost alive, of an almost magical quality. Untouched, lying on the beach, the energy of each wave passing through it, millions of years old, sculptured by sun and storm, tied into an infinity . . . our observation falters but the real definition is there.

But we do touch things, if sometimes indelicately so. Our bodies occupy and displace space in time; to negate our existence from the definitions of reality is to omit a very real influence on the universe's definition of itself. We have been naming and touching things for more than a million years, consciously, and thus have made our subjective presence a part of our universe's reality. We are known in reality. In some small way, our mind's subjective presence is the force that has intruded itself upon that orderly universe interrelationship is describing for us. In our thinking, we have evoked an interplay of forces that had not existed prior to our conscious evolution. If each thing in reality is a perfect representation of the universe there, then our touch is an additional effect, as if from outside, on that representation. To touch that pebble on the beach is to reach back through vast dimensions of space and time incorporated into that pebble at that moment. To caress its smoothness is to trace the dimensions that are its definition at infinity. In that touch we have reached far back; we have communicated through our fingertips the perfect orderliness of the universe imbedded in its vast memory and represented there; with that touch we have also added to its definition a new dimension. Through interrelationship, we have communicated to all infinity that touch and our being behind it and have redefined it within the infinite matrix of our universe. We have lent ourselves to it and given that pebble a new meaning in reality.

What is that meaning? What have we added to reality? If our touch penetrates deep within, into the atoms and the space between the atoms, with our being, then the warmth of our being's energy exchanged for the stone's coldness is a real interaction. It happens in between where the tips of our fingers are in contact with the smooth surface of the stone. But it also happens in the realms of our mind that grants us understanding, and in the dimensions of space that are defining that moment. It is more than just meets the eye; we have done something more. We still do not know what. But we do know that we live, each one of us conscious of our touch, though sometimes reaching out only to blindly and unwittingly destroy that which we are trying to hold. But we see things differently now, no longer as a child. We can be conscious of our touch and name things in a new way. We can become conscious beings, conscious of more than just ourselves. Through our presence in reality, each one of us is tied into that infinite matrix that is our universe. It is a property of our universe that each thing within it is inscribed somehow in its vast network. We are more than just ourselves. We have a meaning out there.

Chapter Four

Each One of Us

EACH ONE OF US had a beginning. In this beginning, we started our development with a small idea, built it up into an infinity, had this infinity, this allnesss, look back upon itself and redefine itself in terms of its own image, each part of itself expressed individually as a small idea, sculpted exactly in its greater image, now a new idea in the universe. This is the process of creation, through interrelationship. In interrelationship, our beginning goes back through our parents and their parents back to the beginning where the first interrelations combined in such a way as to form life. In this beginning, this allness, this image, became man.

Our image connects us with our beginning in all directions. Physically, we are the materialization of the infinity of interrelationship at that point of reality that defines our body. Through the billions of years of life's evolution, through the parentage of our ancestry going back to the formation of first life, we are connected individually to all the forces and circumstances that have created each one of us to exist today. Through time, we have been fashioned painstakingly into the form of our present being; through space, we are connected at every moment of time to the infinite image in the universe that is materialized as the definition of our physical form. Physically, the properties that are our body resemble the properties of the universe that define all things. What distinguishes us is that we live.

It is a property of interrelationship that it can become greater than itself. After all the possible interrelations have been calculated and incorporated into totality, a new image appears. The totality takes on a new value which is the value of its total interrelations plus the value of these interrelations added as a total image. The total image is then interrelated as a new factor of interrelationship, redefining itself through all of its parts into a new image. This process results in a creative force that, in effect, causes the totality to grow continuously, through time. With each growth is a redefinition of all things within the whole. When the redefinition has been completed, the process resumes. In this manner, it is possible for the universe to evolve continuously within itself reflections of its progressively more complex image. At some point, the image becomes complex enough to describe that value we call Life. As a new image, redefining itself in reality as a living organism, in some distant past the universe changed again.

Change took time and through time the universe evolved. With each progressive evolution came a more complex redefinition of both the totality and its reality. The reality that first sustained the simple living organisms became more complex as it accommodated the existence of more complex organisms. Through interrelationship, evolution was as much a factor of the changes in the living organism as it was the reality that defined that organism. With each progressive change in the life form came a gradual redefinition from the now more complex, greater infinity. As the image of totality grew, the life forms that were that totality's most recent materializations grew with it. Each new evolution was a reflection of the newer value added within the matrix of infinity. In each new birth was added, however minutely, that new image. As reality redefined itself and the environment within which the life existed, it changed the organism to adapt itself within the new environment. At the limit, where the change itself is being defined in infinity, much is discarded in favor of that which is to remain and endure. In the end, when the compatibility between reality and the organism is assured, a new life form is born.

The more complex the organism, the more complex the definitions of its environment. With each new evolution the interactions that existed between the living organism and its environment also grew in complexity until such time that the organism would need to register data defining its relationships to its reality. In some rudimentary manner, it began to develop the ability that would enable it to recall experiences, make it more independent of a perfect set of circumstances for survival, and to register this data in its being. At some point, life developed a mind.

The development was always bi-axial in space and time and always registered in the surviving organism. Where the change in evolution was incompatible between reality and the organism, the organism perished, if not immediately, through time. Where the compatibility existed, the organism endured within the matrix of its greater image. The successful organisms passed onto their offspring the elements of their compatibility with the universe. Much was possible, but not all proved feasible. With the survival of certain life forms, the universe grew again. What was stabilized at infinity passed on this definition to their progeny; what was unstable perished.

Each one of us is a descendant of such stable interrelationships. Passed down to us have been all the successful elements of our evolution within the definitions of reality. These definitions have followed us in our development totally and infinitesimally; not a moment was lost nor an experience not registered in our being. Each one of us is the sum total of all the circumstances and experiences that have brought our being from its first living generation to the body from within which we are conscious now. In us registered not only all the characteristics that define our appearance as being human but also all the moments of reality, all the interrelations to infinity, that have brought us here into the present to where we are. We are what and where we are because of what and where was everything else through time. The universe grew with each new definition within itself until its total image created man. At infinity exists for each one of us, individually, a definition in terms of our total image that has materialized us through time into our present being. What is our definition at infinity? In part, it is that definition in us that has registered the data of our being; in part, we are our mind.

The mind is the most recent addition to reality. It is that value that has most recently enlarged the dimensions of totality and given it a still greater value. In addition to all the other values that had been amassed through time, the universe materialized in itself the mind. In man, the most recent arrival in our planet's evolutionary development, has materialized the most recent addition to the totality image of our universe. Our mind is the most comprehensive image of infinity defining us in terms of itself. Individually, each one of us has a mind; at infinity, our mind is the most novel, the most complex image of its totality; for each one of us, our mind is our greatest definition.

In our evolutionary past, when still unconscious, our reality was only slightly different from the realities of other species. If we are to consider that even animals are somewhat conscious, with their ability to like or dislike or to remember past actions that could help them plan an action in the near future, such as procuring food,then we have been somewhat conscious for a long time, even before we appeared as a human specie. But it is doubtful that we were fully conscious beings, conscious of the self, until we could manipulate our consciousness to recall a past in order to construct a future, an awareness of the self and its ultimate death. As unconscious beings, we could not affect reality; we merely lived in it. Reality exerted an effect on us and we responded to it, learning as we went from each new stimulae to the next. Our mind was then no more than a mechanism for registering these events and for helping us respond to new stimulae in terms of what had been registered. It was more like a repository for ideas rather than their creator. New ideas were created but they were still entirely from outside. Evolution was still a very slow process, painstakingly registering in reality each new definition altering the organism's state of being as it changed at infinity. In time, evolution created a mind that could respond to the circumstances of its reality in terms of its own likes and dislikes; the mind could now send back its own definitions. Reality began to register change more quickly as it began to register a novel phenomenon within itself: independent thought. With each additional action and reaction became registered a more complex change growing at a greater rate. Each independent thought became a new value in reality, to infinity, and back, prodding the organism to change again, still faster. When came the conscious mind of man, change became explosive.

For each one of us is represented at infinity a value that is our mind. The mind is the greatest manifestation of all the interrelationships evolved to the present. In our reality, we are the most advanced specie of our planet. We are conscious of the self and in that consciousness are communicating that awareness to all totality that is our universe. The image of that totality defining us is the image that is our mind, individually. If we could each look out into infinity from where we stand and see ourselves out there, we would look past the body and past the animism that is our living mechanism; we would look into that vast interrelationship that defines our mind. We are what and where we are because that is what and where is our mind.

That vast image of reality is not strange to us. In our mind, through time, had been registered all the definitions from reality that had been our history, individually, to the present. Until our mind became conscious, our history was still simply the effect of a universe forming itself; when we gained thought, we joined in with its development and it became a universe forming itself with us. The greater image of our reality, our universe is a definition that had defined our mind, thus, our universe is also the mind that thinks itself in the same way we think of it. Through the successful evolution of our specie, our mechanisms are compatible: we think and we do and the universe thinks and does with us. As we are and do, so it is and does with us, as it has been from the dawn of our history. The fears and hopes, loves and hates, growth and failures, are all part of our past as living beings within our definitions in the universe. What we feel intently, singly, secretly, the innermost reaches of the self, are not unique to us; they already exist as properties of our universe. The universe already responds to all we offer; in its totality are all the characteristics that render us so human. Can it be more than that? We still do not know, but it is reaching us individually.

At birth, we are connected in all directions to the image that defines us in reality. Our interconnectedness through our parents and their parents' parents, all taken totally, has placed us exactly where we are, from birth to now. We are as we are. We can say: "I could have been other, but I am not; I am me. I could have been elsewhere, but I am not; I am here." The image that is the self always occupies its own reality, exactly as it is defined by that reality. In that reality are all the characteristics that are the mind as it is seen by its infinity. The interconnectedness is complete; we live entirely within the circumstances that result from our definition in reality, if untouched.

But to be untouched is an unreality. We are touched at birth, of necessity. The definitions that were our being until birth changed the instant we became handled by another. At birth is for our definition a new beginning in which we are to occupy that space and time that had presented itself as our reality. Our identity had already been formed by the eons of time that had predated our arrival in this reality; we have already been defined by the vast interrelations that were to create our being, from the beginning of time. Once born, we are new and materialize a new identity of reality around us. As a child we already begin to exert influence on this reality with the demands of our mind. As we mature, these demands become more self conscious and more aware of their consequences on our reality. With a more conscious mind, we develop the ability to exert the self in our reality and thus effect change in it. Our reality takes on more of the influences of the self and rematerializes itself accordingly, in response to how that self is in relation to its identity at infinity. Depending upon how each thought and deed is received at infinity will result how reality will materialize in our environment in response to the self. Thus, the mind materializes its own reality, from birth, and the environment of its existence is always a reflection of itself.

We think and thus we are. At its simplest, with the mind still unconscious, this relationship is in direct proportion to the stimulae from the mind's identity at infinity. Untouched our thought and our being are equal. When conscious, the relations become more complex, a greater interplay of forces between the mind as it is at the self and as it is at its greater identity. The relationship between our mind and being becomes more a communication between the mind and its materialization at infinity, its environment. Conscious, capable of creating thought, it is also capable of upsetting that delicate balance that exists in its real identity. Touched, an identity can become other than the self.

If the mind's expressions are real, they become accepted by reality; if false, they become rejected. If we seek increasing acceptance, then the universe thinks with us and the self grows within its greater environment into a still greater consciousness. If we continuously court rejection, we are frustrated into an existence less conscious and continuously more calloused. More conscious, the universe grows more conscious of us; more calloused, while we live, we become more hardened against its definitions of our reality.

Each one of us from the beginning of our evolution, and later from our beginning at birth, is both affected by and affects the identity that defines us at infinity. In that definition is recorded all that has contributed to our being and to our self awareness as conscious beings. We are the product of the universe's growth, its creation within iself; having a mind, conscious, we are now also contributors to this growth with our own creations. As we create, the universe creates with us, in its image. Untouched, each one of us is an image that is being created by the universe, out there. In that image is the consciousness of our new identity. In our being, in our mind, from the dawn of our creation to now, at each moment of time, are the mechanisms that are compatible with our identity out there.

Chapter Five

Each One of Us, Conscious

EACH ONE OF US, conscious, has a mind. That mind is our identity in the universe. Conscious of our identity, we are more.

Each one of us is surrounded by a unique reality. In our individual environment is the definition and materialization of our mind as it is seen by infinity. In that mind are all the elements of our identity at infinity. At infinity, each one of us is unique.

We still are not conscious of all the elements of our mind that are our identity. Our identity is far greater than what we think of in our personal being when we think "I am." Our consciousness allows us to think that we have a mind; it still does not allow us to think, entirely, what that mind is. But now we can know that the mind is greater than the product of its consciousness. In our mind is an identity greater than the one we know; but this identity materializes our reality and to this greater identity we must defer. Conscious, if we are to more fully occupy our space in time, our reality, we must become more conscious of that identity; conscious of that identity, we become more.

The reality that defines us in our identity envelops us completely. Unconscious, this is always so, since the organism only responds and is thus always occupying its defined space in time; it cannot stray. Conscious, however, we have the ability to choose action through which we may momentarily step from our identity. In each thought, each decision, each action, is the potential to either work in the direction of our identity or against it. Both have their purpose and their results. When we work with it, our environmennt is enriched with the materialization of our success and our being is elevated within our identity; when we work against it, our reality opposes us and leaves its imprint on our being. Such is the mechanism that allows us the freedom of mobility within the medium of our reality. But this freedom must be exercised within the limits of our responsibility to ourselves, that which defines us in the image of our identity. Earnest, ourselves, as defined by our identity, we move with our reality; false, we rest opposed.

There is, in effect, a true way and a false way to work within our reality. How reality judges us is how that value of truth or falsehood, in relation to the definition that is ourselves, is materialized both in ourselves and in the circumstances that engulf us in our environmment. These judgements are not in relation to a predetermined or predestined set of rules of what is right and wrong; they are in relation to that personal value of what we are as an identity, our true selves, as defined by the image that is our personality at infinity. With our minds we can now know the reason for our being conscious of our identity. Conscious, we can then choose to either work with or against ourselves.

We have a definition in our reality. That definition is the state of being of the rest of the universe that envelops us completely. It is a characteristic of that definition that it has a value at infinity which renders us human. That value at infinity is our identity. As we live, so our reality, through our definition, materializes that value that is our identity. Each one of us, in our daily existence, reflects that identity that is the image of our greater self. We can imagine that greater self as a kind of super life-support mechanism through an otherwise inhospitable environment. But it is more than just a mechanism; in it is reflected that value at its totality that to now we would have expected to find only in the human mind: consciousness. It is our mind "out there."

In our mind are all the values that characterize our identity, that which renders us human. Our greater self, our mind, has the impetus to adorn us, to make us more beautiful, to clothe us lavishly or simply, to help us express ourselves poetically or directly, melodiously, kindly, lovingly, subjectively or objectively, to make us more than mere organisms that efficiently process their environment from one form to another. We are creations given the ability to create; thus we materialize around us visions of those forms and forces that are creating us. The mind has the desire, the need, to express itself beautifully, richly. Those are the characteristics of our human identity; at infinity, sincerely, we are all those things; in our immediate reality, in the definitions that surround us, we are reflections of that level of sincerity that marks the level of our humanness within our identity.

As each one of us lives, we create in our surroundings a reflection of our inner and outer being. As a man or a woman lives, he or she naturally leaves a mark of his or her being on all things he or she comes in contact with. We cannot help ourselves; our presence is a phenomenon of our being in the universe. The universe has described us to look and be as we are; this being is then reflected in ourselves and in our immediate environment. If we are smooth, soft to the touch, gentle, then that is our definition within our reality. If we are hard, sinuous, resolute, then our reality is characterized by that definition. As we live, we leave behind us a trail of our deeds; we materialize our being in the way we fashion matter. Our conscious mind gives us the power to change our reality and in how we change it is always reflected as our being in our identity. We strive naturally to surround ourselves with what pleases us. As we succeed in this endeavor, we materialize in our environment more of those characteristics that are our mind.

The body, the dimensions of our inner universe, forms itself in relation to our being as it materializes within the space-time dimensions of interrelationship. It is born with our identity already marked on it. As to how the mind molds it in its development from birth, we still cannot know; but we can know that through its interconnectedness to its identity at infinity, it is formed in reflection to its greater image out there. If it is pleasing to the eye and the touch, then those are its characteristics at infinity in relation to the observer who finds it so. These are natural reactions to natural definitions of matter as defined by infinity. Its identity will keep this definition as long as it can before it weakens with age and begins to lose its original form.

The body strives to occupy its form in identity as long as it can, preserving its youthful energy until it surrenders itself to the serenity of age. It will strive to leave behind as much as it can with its touch, its mark on reality. Where it walks, touches, gazes upon, is; these are all registered in the greater image of its reality, itself left behind in its greater self. Where it feels, thinks, dreams, desires, are even finer influences on the matrix of reality than where the skin comes in contact with the universe. These are fine, delicate influences that are felt subtly throughout reality but which, in the manner of infinity, are registered back to their origin. The finer the influence, the more beautiful, gentle, kindly that touch, the more graceful the reality that surrounds that being. From inside is projected our universe outside, as defined by the outside that is forming us. It is an interrelationship of beings, our inner and outer being, and in both is registered our identity.

We project ourselves naturally, as simply as we breathe and live. As we adorn ourselves naturally, in our dress, our care, thought, our manner, and as we adorn our reality, our homes, our work, our gardens, our children, our communities; all are directed by our being as it becomes real in between the forces of our mind and its greater identity. Our mind strives to express itself in a way that will make it more itself in reality. We tend to fashion reality in our image; our image out there tends to fashion it with our mind; together they create what is visible to us as a single living being. We do not create this world solely from within ourselves; all we do is supported by the state of being of everything else in its totality projecting in all directions from our being. As we do things, the rest of the universe does things with us. That is how we project ourselves into reality, with the help of our greater being. That we strive to project beautifully is a characteristic of our human identity as it had been defined in our mind. We touch with our mind beautifully. It is the essence of our being. Our existence is a natural covenant between ourselves and our universal identity. They are in agreement, working together, adorning one another in each other's beauty, a reflection of one another. Untouched, that is our natural state of being in the universe. But we are not untouched.

Even in nature, even primitive, man adorns himself. It is a natural characteristic of our human mind to improve on the unconscious. Nature, in its wilderness state, has a strong appeal to us. Its forms, its movements, are pleasing to us. We seem to gain a source of strength from our communion with it. Yet, our conscious mind finds the need to express its individuality. We take possession of things, work them, move them, mark them with our personality, name them, and then leave them in our immediate environment to characterize our being. We identify things, in effect, lend them our identity. As we become more sophisticated in our command of reality, our environment becomes more sophisticated with us. The naming and marking of things become more formalized and we have laws; our working of the environment becomes more mechanized, more an extension of our understanding of reality, and we have our modern technology. The more conscious we become, the more we take possession of our reality and the more we leave our identity in it. We fashion it in our greater image, superimposing the conscious mind on the establishment of the unconscious. The unconscious was communal, free, developed; the conscious, even semi-conscious, is territorial, independent, individual, developing. Having both characteristics in our being, we have a deep affinity for both realities, the primitive and the sophisticated. As we project our identity on our reality, we are drawn into it, but we cannot abandon our past. We are drawn to nature and its ways and though we are striking out into our new independence, our development in our future consciousness, we need not do so at the sacrifice of that which is natural to us, that which had brought us here.

We may sometimes feel like intruders on nature, with our conscious minds. But nature is more than merely the wilderness. Nature works unconscious, perhaps conscious only at its totalities, whereas man is conscious individually as a materialization of those totalities. Perhaps we called those totalities gods once, but now they are only definitions of greater images within the interrelationship of reality. We are only intruders when we are not conscious of our consciousness within nature. Conscious, we left our primitive past and occupied a new identity in the universe, a new responsibility. We are the elements of nature's latest development, we are the latest evolution, and in that evolution is the responsibility to add our humanness to our natural order. More human, more beautiful, we will adorn our environment more in our image as nature adorns us more with the beauty of its creations. We are not intruders in the wilderness; conscious, the wilderness is our trust. It is a trust to mold in our world a newer, more human, more graceful reality. If nature is a living reality, conscious at its greater totality; if it had carried us in our development until we reached consciousness; then it is now our responsibility to carry its living totality into our future reality. In it is our beginning.

The individual conscious mind incorporates in itself all the values that render it more beautiful, more human. The things that we create to adorn ourselves, the bodies that materialize our inner being, the nature around us, and the total reality we occupy in the universe are all sensitive to the mind's definition of its identity, its outer being. The mind and its identity are a mirror image of one another and between them is the creation we call Life. Conscious, that Life is the life of an individual. As an individual, it is the responsibility of that life to treat its reality with the responsibility with which we had been entrusted. It is our role in life to express ourselves within our identity and to thus create the world in our greater image. Conscious, we do this naturally, with our minds and bodies, as we adorn nature and the world around us.

Within our immediate reach is how we live, how we look, how we treat things; at a distance is our social environment, our art, our thought, our disposition toward ourselves and others, our planet and our universe. It is the mark of a mature race that approaches these with grace and reverence; it is in the conscious mind that this grace is individually evident. That consciousness is the mark of a more human race versus one that is still primitive, still unevolved.

A man or a woman can be more beautiful, completely. Conscious, it becomes so much more than merely a body and mind aware of itself; it becomes a mind and body aware of itself with its reality. Conscious, it has the power to move the universe in its greater image and become more as that image has already molded it. It is greater, richer, gentler, fuller, deeper, warmer, more beautiful than before. Its dimensions reach deeper into those values of infinity that are molding it into its greater image. Human, we have a richer, more beautiful woman, man. More beautiful, we then project a more human, more sensitive reality. As we live and occupy our space in time in reality, our identity, we move with it as it develops us in our image. We create and it creates with us. We prolong ourselves in it and in this prolonging we define our environment in this greater image. We are entrusted with this creation but it demands that we lend ourselves to it to enrich it. We are the builders of our universe, but we can grow only by growing in beauty. More beautiful, more human, the universe grows more beautiful with us.

We have a trust and that trust is that we become more conscious, more tender in our touch. We are adding to our universe, but we cannot add when we are destroying it. We adorn ourselves, but the universe adorns us with our greater being; we cannot do this when we are working against it. More conscious, our reality becomes more adorned with our more human identity. The trust is one of tenderness, of grace, of reverence for what we have created; it was created by our mind from its inner being, but also with its being outside. If something is beautiful to us, it is because they have worked jointly, in agreement. We have been adorned and, conscious, we adorn in return. As we sculpture the world around us in a more pleasing image, we move all those forces and circumstances, those patterns of interrelationship that define us, in a way that is guided by the mind and its identity. When we succeed in joining the values of our mind, our being, with the values that are our identity out there, when the tension that exists between them is fulfilled with our creation, we create or discover a thing of beauty. It pleases us, it moves us; it moves us closer to those values that are our identity. More in our identity, more conscious still, we create still more richly. Our reality reflects even more closely our identity and we become even more adorned in it. More conscious, more our identity, the universe entrusts itself more to us. We are creators, not destroyers; creators, the universe creates with us.

It benefits man to seek his identity; — "Who are we?" — it is how we become more human. As we are free to create, to adorn ourselves, as we live, we come closer with our mind to that value of infinity that defines our identity. As we create more in agreement with that identity, we become more within it. Conscious, aware of ourselves in our greater identity, we become more within the universe as it is making us. As we grow, more mature, more upright and graceful within the body that adorns us, we project more of ourselves with our touch and our being into the natural order that surrounds us. Having a mind, conscious of that mind, conscious within its identity, we lend that mind into the reality of our universe. We become known and accepted throughout infinity, and thus we become elevated in how the rest of the universe perceives us. More beautiful, unblemished, softer, clearer of spirit and more elegant in thought, we approach more closely in our body and our surroundings that value that is the spirit of our identity. More in our identity, we become more man. Untouched, we are so much more. Our pursuit of beauty in consciousness is the essence of our development and we could dwell on it here forever. But we are not yet at liberty to do so. We are not untouched. There are other pressing demands of reality that we must cover in our development before we can return to our pursuit of our greater being. It will not be difficult, but there are conditions that must be still fulfilled before we can approach ourselves as truly human. To gain our identity, we must first become ourselves.

Chapter Six

We Can Choose

W E CAN CHOOSE OUR ACTION in response to
each circumstance. We do not have to choose and can
go on in life creatively inactive, much as one would being
another's ward; or we can exercise the other extreme option
and take a final act, to die. As conscious human beings, we have
the power to decide whether we act or do not act, whether
we live or die. These are severe choices, but they serve to illustrate the importance of our ability to choose.

Our choices have real consequences. Man is more in his mind
than the sum total of his experiences; man has an independent
will. We have in us, each one of us individually, a distinct and
separate will that grants us the power to create, to change the
universe in our real image. It is in our identity in the universe
that our definition as human beings is endowed with that
creative ability. It is demanded of us the same as we demand
our freedom to exercise our will. Our freedom to choose is an
integral part of our human existence.

How we choose in the world is how we occupy our identity.
We can either seek to better occupy our identity, in effect,
become more ourselves, or we can choose to become other
than the self. In one, our will works to make us more conscious,
more attuned to our greater reality; in the other we drive our
will with no regard for any greater reality beyond ourselves.
One works with the universe and, consequently, has the
universe working with it. The other isolates itself from the
universe and its effect on the mind. In itself, this choice whether
or not to ignore the universe and one's greater identity would
not be of great consequence since it only affects the self and

37

the erroneous choices would be corrected within that mind's greater reality. The damage becomes more apparent, however, and takes on greater importance when the mind, choosing to not be in agreement with its universe, drives its will into the reality of another. Identities clash, personalities come in conflict, and, if the aggressor is successful in his trespass, the victim is forced from his space-time position within his reality, his greater identity. Trespassed, that mind is no longer free to be itself, disconnected from its definition within reality, outside its identity. Thus violated, wounded, its agony defines it in reality as a victim of an aggression; if the aggression persists, it is the victim of its identity's captivity resulting in a loss of personal freedom.

Our desire for freedom, our need for it, is a real phenomenon. We must be free to occupy our space in time. It is a freedom we demand individually and, when we are forced from it, we are forced from the definition at infinity that is our identity. Thus forced, we become filled with tension, we hurt psychically, become filled with unhappiness and we feel the need to come back into ourselves. This happens naturally, and to be pushed in the wrong direction, against our will, against our agreement, triggers in us the response to push back. This trespass can be corrected effortlessly and, in a civilized society, the aggressor realizes his or her trespass and apologizes or otherwise makes amends. Unfortunately, as we grow intellectually more sophisticated, we invent ways to legitimize this trespass. This too has happened naturally, but now we can choose. We do not have to accept a trespass.

Our history had been replete with the trespass against our will; it had been the nature of things. Chiefs demanded a communal allegiance within the tribe and punished any who disobeyed. Kings ruled their subjects as they willed and the subjects accepted this rule by personality. Only recently had government by law rather than rule by personality taken hold in our social consciousness and become responsible for the successes of modern times. These successes can be witnessed, if only generally, by the contrast between the social advancements and economic progressiveness within societies ruled predominantly by law as opposed to the backwardness of those still ruled by the cult of personality. Where the arbitrariness of will had been replaced by the, if still imperfect, impartiality of law, the trespasses under which the former subjects were forced to live increasingly vanished. Our more modern constitutional governments, when properly safeguarded by individuals, tend in their intent away from arbitrary force against the individual. Laws are our safeguard, in principle, against the arbitrariness of tyranny. However, today, to be pushed against our will, when it is not the result of a trespass by another individual, can be the result of a law that forces us to be other than ourselves. Laws can protect us from the trespass of another; they can protect our property, our rights as individuals, our lives, our agreements as expressed by contract. But can they protect us against themselves? Even if inadvertently, even without design, a law can be made such as to trespass against us, to force a body to be elsewhere and otherwise than it will, against its agreement, the mind other than itself. When the conflict or trespass was but an act of individual wills, the solution was simple: if not otherwise resolved, combat. Our history is a long trial by combat. However, when the trespass becomes one of ideology and social structures that seek to dominate the mind, the will, then the resolution becomes more complex. Combat exhausts itself and the tyranny, even if inadvertent, of the mind triumphs.

Thus, laws cannot always protect us as individuals. They cannot protect us if they are written such that they negate the value of our individuality. If the basis for proper action is the benefit of the whole group rather than the benefit of any one individual, then the laws that guide that action are such that the right of the individual are subordinated to the right of the group. But the benefit of the group had always been favored historically, even if that group's welfare was embodied in the representation of one individual, i.e., the king. Only the image of the representation of the group has changed, not the underlying principle of social order. Where the social order of the group had been dominated by one ruler and his counselors, the new social order is dominated by a constitution of laws and their elected representatives. There is no harm in either system, save when either system strives to subordinate the right of the individual to the right of the group. Then, the individual ceases to be sheltered from trespass and falls victim to aggression. This aggression need not be intentional, it need not select anyone individual over another; it is merely an arbitrary trespass against individuals who wish to be themselves. So trespassed, the individual is forced from his reality by the very laws that were designed to protect him or her in their society.

Nothing need change. The benefit of the group can remain favored as it had historically. The laws that defined the social order and its government need not change; nor is there the need to change the social structure. It is actually in the interest of the group to become elevated within itself, as it desires. But the social group cannot elevate itself, in effect favor itself, if it does so at the expense of the individual. It need not subordinate the individual to the group and thus force it to be less than itself. It is possible to establish a group that is defined by an assembly of free individuals, free before the law to be themselves, with no detriment to the society as a whole. We can now do that with our understanding of the mind in interrelationship with its presence in its identity. Our society can be made to be responsive to the demands of that new man, new woman, without subordinating them to the force of the group. They have conscious minds, they choose their actions, they have a will and an identity; they are individually their own force. It can be accomplished without changing the social structure. It is a very small step from where we are now to where our social order is run on the principle of universal order. It was only a small step from being unconscious to consciousness. It is now but an imperceptible step that takes us from being subjects of our social group to being subjects within a social structure subject to a greater universal order. It is our right to choose, our right to consciousness.

Unfree, we do not have the right to choose. We are unable to select whether we will occupy more closely our identity or whether we will force ourselves from it. The choice is often made for us, by the demands of the state of being of all the others, the group. Their claim is to self preservation and advancement; the individual's aim is to contribute. But if this contribution is other than how dictated by the group, the individual is considered to be the one in error and to be corrected. When the greatest legitimacy in our mind is the legitimacy of the group, it being the greatest materialization and representation of order, of intellect, of rationality, then it is justifiable that the individual who has strayed from the dictates of the group is in opposition to the greatest materialization of order. It is, as such, unjustifiable to be opposed to the group. If there are faults within the group and these faults cause conflict, then they can be, must be, resolved and there can be room for debate and improvements. But that is the trap: once the social order has been perfected, argued complete and self justified, there can be no justifiable dissention. To dissent, under those conditions, is equivalent to denying the legitimacy of human intellect as supreme. It is to be unjustifiably rebellious, to be socially ungrateful or, worse, to be irrational or insane. In that closed world, there is no outside legitimacy to order; it is entirely self contained within the structure engineered by the mind. Thus, the right to choice becomes a false issue, since to refuse to choose along with the group, when fully legitimized in the public mind, becomes an irrational act. The system is self enclosed around the group allowing freedom within its structure. Freedom, then, is not a self defined, unalienable, right to be oneself. Instead, it is defined by the rules the individual must obey within his role in that social order. Whether or not that role is his or her real identity is a meaningless question. Where no greater intellect is acknowledged than that intellect manifest in man and applied to social planning; where there is no greater order than the, albeit if still imperfect, social order; then the role of the individual, in social terms, is his identity. If he or she rebels, then they have been improperly socially programmed; they are insane. To our new consciousness, that is unfreedom.

While that society is still imperfect, choices do exist. The choices are in how to make that society more complete, to tighten its grip on its own activities and make it more productive. Then, all intellectual efforts are applied to this end. To choose otherwise is to undo what the group is doing, to go against order, to cause social conflict. There is no room for the soul to create; the mind must create in its socio-political context. To rebel, to stray is to become outcast, banned. The trap closes as the society approaches its self-defined ideal of perfection. The closer it approaches its own perceived utopia, the less free become the individuals within it. In reality, this probably cannot actually happen because society is unable to approach that utopia closely enough without creating immense social cost at the expense of falling productivity and human vitality. But, for a theoretical purpose, a utopian perfection is the intent that drives the group in this social venture. All efforts are thus channelled not to better the individual but to better conceive and execute the perfect group. All choices are geared to that end and all efforts are gauged in terms of those results. The success of the individual is then judged in terms of the success of the group and that individual's contribution to it.

It is curious that such a will, such drive for human progress, can actually be its greatest impediment to it. Our thinking, our philosophy was based on a closed universe; it revolved entirely about the supremacy of our human intellect. We could not see beyond it and were closing in our society. If sometimes brilliant, we were, regrettably, building ourselves a social trap. There is no blame; we were not evil in our intent. We simply acted on how we perceived our universe and ourselves in it. It was a universe without God; we were individuals without a soul. We did not see that we were backing ourselves into a box that would trap us and take away our freedom. The harder we worked, the more brilliant our dissertation on social order, the tighter we backed ourselves into it. We had built ourselves a trap and we were expertly trapping ourselves in it. We blocked out the need for self expression and limited our right to be ourselves. We could not even rebel, for to do so would be insane, to be incapacitated by our reason and convictions. And yet, in spite of it all, we continually rebelled. Weakly, guiltily, without conviction, madly, we chose to rebel against an increasingly orderly society in favor of a more natural, freer one. We chose freedom, even when all our intellectual senses marshalled against us, even when there was real danger of social reprisal. The more sensitive, the more conscious and beautiful the mind the more it craved this freedom from the established order. But it was trapped. The mind sought freedom, but it found regression, simplicity, primitivism, communalism. But it does not have to regress to be free. We now have the tools with which to construct, philosophically, the foundations for a social order based on the identity of the individual, based on his freedom to choose. From these foundations we can implement the small changes that will propel our society beyond its present impasse. As conscious minds, we have the ability to choose our freedom. Each one of us can be free as an individual to occupy his or her space in time within the definitions of his or her identity. What at first appears to be incredibly complex can be, in its final analysis, incredibly simple. We can be free individually within society and watch it become elevated rather than, as one would still suppose, have it collapse in anarchy. Freedom is the power of the mind working with its universal order working with itself. We can apply this power to our social order and create progress beyond the impasse of the present.

We have progressed from a social order that was based predominantly on our need to survive the natural environment of this planet to a social order creating its own environment based on or understanding of reality. We had gone from a primitive, semi-conscious mind to one that became conscious of the self. Conscious, our community evolved from the primitive simplicity of the cave society to the sophisticated complexity of our modern social order. As we choose to become more conscious, as our human needs increase, we develop a more advanced and abundant society which offers us both greater personal freedom and greater material comfort. We have been doing it naturally almost by instinct, led by a hope, a wish, rather than an accepted premise; our drive for freedom was something we really wanted rather than something that, to our minds, was inherently defined in us. We believed that feeling, that want, and succeeded in building on our planet the most advanced socio-economic structure of our history. It was an accomplishment; in such a short time, its development was spectacular, explosive; it was because we believed in something that is true. We are free individually; we are really free. Our definition in our universe as free, conscious minds, free to seek to occupy our greater identity in the universe, is the definition of our freedom. We have progressed, materially, to this level; we can now be free from want to pursue spiritually our identity and find ourselves out there. We need not regress materially to advance spiritually. We have a real need for our material world to adorn ourselves; we need our comfort to experience our freedom as conscious, human beings. But that is the choice we must make. It is a simple choice; once made, what follows is no more difficult than what had been our common experience as a human society. It is simply the right to choose to be ourselves.

Chapter Seven
Having a Mind, We Have a Right

HAVING A MIND, we have a right to choose to be
ourselves. Conscious, our mind has an identity that defines
that consciousness in its greater image; our consciousness is
our corporeal definition of our mind out there. We have a value
within our greater identity; it is how the universe sees us within
our space and time in terms of itself, at infinity. That value is
that, having a mind in a corporeal body, we have the right to
be ourselves with our mind and body within our greater identity; that we be free to be ourselves in our identity.

The choice is ours. We do not, individually, have to accept
this freedom. We may choose to live under conditions that
to another may describe us as unfree. We may wish to be
taken care of, to pay the price of freedom for the comfort
of a known security, or we may simply wish to always surrender
when threatened rather than risking resistance. What may be
freedom to one person may not be appealing to another. But
the choice is ours and we must have the right to choose, to
choose which way of life is in agreement with our personality,
our identity.

The right to choose to be ourselves is pivotal to our existence
as human beings with a conscious mind. We have a corporeal
value in infinity because that is how is defined the mind, there;
because we have a mind and that mind chooses the right to
seek itself within its own identity, it must have the right to
choose how and where its corporeal being will be. How it
chooses this will reflect how well it occupies its identity within
infinity. In how well it occupies its identity will be reflected how
well the universe works with it and is adorned by its presence.

Individual freedom is the next level of consciousness in our human developments; it is the realization of our humanness in terms of out there.

How do we approach this freedom? When we were unconscious as beings, there was no freedom; we simply were done to by reality. Conscious, our freedom became more important in terms of the level of our consciousness. The more conscious, the greater the need for the right to choose in response to how we were done to by others. Is it possible, then, to gauge how developed is a person's level of consciousness in terms of how demanding that mind is of its individual freedom? In part, yes. But there is more. A demand for freedom, to be justified in terms of the definitions we have given to freedom, the right to choose, must be in terms of its consciousness within its universal order. When a mind can accept no greater reality than its own, then it is not yet ready to choose its freedom. Freedom given to that mind is dangerous. It must be given to the mind when the mind demands it in terms of its greater identity, responsibly, consciously, not indiscriminately. To give out freedom wholesale would be equivalent to giving total freedom to children before they reach mental maturity. Until they can become cognizant of the consequences of their actions, and become responsible to these consequences, then total freedom would be wholesale chaos. This is not what we would inflict on our young planet. There is a method to freedom, in terms of our greater reality. How we choose to approach it is the level of our social maturity.

It is conceivable that we could grant freedom to each person who demanded it, who stood before a social tribunal and demanded that, being a conscious mind and aware of its greater universe, he or she be granted the freedom to find oneself in terms of one's identity. As each youth or adult approached that level of maturity, they would be subject to tests and interviews until such time that they be granted a special license that would declare them free. It is conceivable, since stranger things have happened. But it is absurd, cumbersome, complex, ambiguous, arbitrary, and too subjective to approach in reality its intent. We simply do not know, nor can know, the identity of another to such a degree as to be able to pass judgment on it. It is forever a personal communications between the person and his or her mind at infinity. Same as no two persons can occupy the same space at the same time, no two persons can ever experience the same identity. It is simply impractical, if not outright false, to attempt to define another person's right to freedom. But, through interrelationship, things can now have their own definitions; things can be judged in relation to themselves. There is a simpler way to approach freedom.

There are two ways for a person to gain freedom. The first is for a person to declare himself free and then fight anyone who dares to disagree with him. That is barbaric and, in a civilized world, absurd. There is no way to gauge whether or not that person is actually free in terms of his greater identity. Such a declaration of freedom is entirely within that person's definition of himself, void of any apparent knowledge of or responsibility to the consequences of his actions within a greater reality. Thus, the declaration, as such, is not justified, in and of itself, in terms of that person's greater identity. There will be times when that person's declaration of freedom is true and times when it is not, but these will be judged by that person's subsequent actions and not by his or her declaration. The second way to approach freedom is in terms of greater identities, in terms of their interactions. Because interactions of identities are a social act, then the second way is in terms of society. We can insure freedom either individually, through combat, or socially, through a social order, one that is organized around a concept of freedom and which strives to preserve this freedom for each individual. Ultimately, historically, we tend to gravitate towards the social solution rather than the anarchic one. Now, society can be based on a free social order, one that insures the right of the individual to choose to be himself or herself.

A social order can be based on a tyranny, where the society is organized for the benefit of one individual or group and where subjects of this society are ruled through coercion and fear of coercion. Or, a social order can be based on liberty, where individuals are free to seek agreement among themselves and are protected from trespass against these agreements by the social order. The former is a government of disagreement and needs force to keep its subjects in submission. The latter uses force only to enforce its social contract that, through its laws, agreements between individuals be protected. One is government by force, the other is by contract. One seeks to dominate the individual; the other is to set him free.

If a social order is organized in opposition to anarchy, that is, it is organized by contract to shelter the individual from constant aggression or combat to insure one's freedom, then it is organized for the benefit of the individual. It shelters the individual's right to safety within that order. If, instead, a tyranny sets the individual in constant friction, both with the social order as well as with other individuals, then there is no advantage to joining such an order, except in the case where an individual wishes to join the aggressors and thus gain from coercion. But a free individual cannot exist in such a tyranny as a conscious mind except secretly, as with a double identity. For an organization of free individuals to be viable, it must be ordered on the premise that the individual will have more to gain, to suffer less, in a society free from coercion rather than in an anarchic state, as in the wilderness. Taken to its rational conclusion, a society that is consistent with an order of conscious and free individuals is one that is least coercive and that insures each individual's right to his or her freedom from coercion.

Such a society, if it already exists, would be expected to have an existing tradition of human rights. One such basic right, in addition to some protection for personal property and personal safety, would be a law that protects the individual from coercion through a writ of "habeas corpus." It would be a requirement that a person "have a body" in his or her defense. It would also follow that such a society would also be in a position to incorporate into its social structure the equivalent of a writ of "habeas mentem" requiring that a person also be present "in the mind" in his or her defense. A social order based on the idea of "habeas mentem" would be one where coercion would be still further restricted. It would be a society where a conscious mind, before the law, each individual, would have the right to defend the self from trespass with his or her presence not only in the body but also in the mind. It would be a society of Habeas Mentem.

By our definition, having a mind, a person's presence has a greater meaning. A person, conscious, is present in more than merely the body; conscious, the person is also present in his or her identity at infinity. "Who am I?" is a presence that demands an identity and who is violated when trespassed. If the "habeas corpus" is a body present, "habeas mentem" is the presence of the mind. That mind is now greater than merely the person's intellect; it is also his or her total being. If a society believes that its greatest legitimacy to order is embodied in its intellect, in its rational brain, then it is a society not yet ready for Habeas Mentem. If the product of rational thought as embodied in social laws and socially desirable engineering are perceived to be the highest good, then the level of consciousness has not yet been achieved to allow the existence of Habeas Mentem, for an idea of total being, greater identity, is still a meaningless concept. For a society to accept Habeas Mentem, as it had already accepted "habeas corpus," the society must also accept that the universal order is a greater legitimacy to order, even social order, than merely human law. This is a giant leap of consciousness, but in order for Habeas Mentem to be understood, it is but a rational conclusion of the idea of an infinite interrelationship. It is a test for the individual.

For the individual to gain the right to be in the self, in "who am I?", before the still greater tribunal of his or her greater identity, he or she must choose to be the self. A mind free from trespass, free to choose to be itself in the universe, is a mind that is present in both the body and the mind and, through Habeas Mentem, is free to stand in is own defense. If untouched, untrespassed, not forced against its will and agreement, in agreement with its greater identity, that mind is its own representative in its body. If trespassed, however, coerced to be other than the self, then it is pushed out of is reality and no longer in agreement with its greater identity. When trespassed, we do not "have the mind," are not occupying our greater identity and thus are not representative of our will. We are then without Habeas Mentem, unsheltered, unprotected by social law, as in the wilderness. When victims of such trespass, we cannot be the aggressors. Outside the mind, as victims, we may not be held before a tribunal, for then we are not in the "body" of the mind. In a society of Habeas Mentem, this basic protection, to be in the mind, is the first right to help us regain our identity. We are victims until we have the mind. Until then, we have the right to stand only before the greater tribunal of our personal identity's reality at infinity.

In Habeas Mentem, we are judged first by the greater tribunal of our mind in the universe. That greater tribunal is our conscious mind as it is expressed in our presence of mind and in our conscience. If, within the mind, we are found to be guilty of force against another, guilty of trespass, then we are brought forth before the tribunal of our social laws and judged within their context. If, then, it is judged that we are innocent and that, rather, it is we who had been trespassed against, that it is we who are not in the mind; then we are to be free to choose to seek our personal freedom from this trespass and to reestablish our identity, to again become ourselves. We can lay claim to Habeas Mentem and declare that, victims of trespass, forced against our agreement, we have the right to be free from further trespass, that we be unhindered by aggression from seeking our greater, more conscious identity; that we have the right to choose to be ourselves as we consciously will. If coerced, forced from that right, we do not have the freedom of the mind to stand in its own defense. However, if guilty of that force and thus judged by the social law that it is we who had forced another from identity, that we are the aggressor, then we do not have the right to seek protection by laying claim to the right that grants us the freedom to have our mind. Then, unprotected by Habeas Mentem, the aggressor becomes subject to the judgments and punishments of social laws. Habeas Mentem protects only while innocent; it offers no shelter when guilty. Without the right to the mind through our act of force against another, through our aggression, we then forfeit the protection of Habeas Mentem until we regain the right to our identity. Until then, we are entirely subject to the human laws of our society.

Habeas Mentem is not above social law, unless such law acts to trespass on a mind's agreement with its reality; then, it is the mind's first protection. Habeas Mentem is not above the individual, unless that individual acts to trespass against another;

then, it is beyond the reach of he who is the aggressor. Without Habeas Mentem, the conscious mind guilty of willful aggression is then brought before the social tribunal without the right to the mind until there is proper restitution to the victim. Thus, we are free to be ourselves, within Habeas Mentem, free to have our mind and to seek our identity, until proven guilty of trespass against the identity of another. But, proven guilty, we are then without the mind, we have strayed within our identity by causing damage to the identity of another, and until forgiven are entirely within confinements imposed on us by the victim or his social representatives. Without Habeas Mentem, we are then free only to the extent allowed us by the laws of our society.

We are free to seek agreements with our minds, both with our reality and the reality of others; in disagreement, we are protected from trespass by Habeas Mentem. A society that incorporates in its laws the principle of Habeas Mentem recognizes the greater universal value of the individual within his mind, conscious, and desirous of his or her freedom. Thus, it is a society that protects that individual from coercion that would force him or her from their identity, their mind. To force a person against his or her conscience, against the dictates of the mind, is to force that individual to act in a way that is contrary to that person's definition in his or her identity. In a free society, each individual is free to seek his or her consciousness as dictated by his or her conscience. Provided that seeking does not trespass on the reality of another, provided that it does not go against another's agreement, then to seek to become the self in reality is a freedom wholly safeguarded by the principle of Habeas Mentem. Each individual in such a society is then free to materialize his or her value at infinity in the environment of their social reality. It is a society that shelters freedom and which does not allow tyranny, even the potential tyranny of its own laws. It protects the right of the individual by safeguarding the laws of its social contract from forcing individuals against their agreement with their mind and the minds of others. A free society recognizes each individual free until proven guilty of trespass; if victim of such a trespass, then the individual does not have the mind in his or her defense against alleged accusations. Simply, a mind not guilty of coercion is guilty of no crime.

The society based on the principle of Habeas Mentem is a society of free human beings. It is a free society, free from tyranny and free for each individual to seek himself or herself in their identity in the universe. More conscious of reality, reality more conscious of them, they materialize in their environment their labors as reflected by a universe that is laboring with them. It is a simple value, freedom, and yet it can become the most powerful impetus and generator of social advancement. In a world free from coercion, each person seeks his or her rightful place in the order of the universe and is free to materialize the value of that rightful definition, with all its near infinite ramifications, in their immediate reality. They become creative, productive; it is a new world, one never seen before in its entirety; but it is the world that is inherent in the next stage of our human development. A consciously free society, aware of its freedom, is the foundation for the development of future man.

It is not so strange, this concept of Habeas Mentem. It is simply a method, a principle, with which to gauge whether an individual had gained, in his mind, freedom. Freedom is a gift of the conscious mind; we are not free to choose until we are conscious that we wish to choose and then how we do choose. Then, we are free from trespass and tyranny against our person, for then we are free to choose our greater individual identity within the safeguards of our society. We are free from the perpetual need for combat to insure our freedom; a social order based on Habeas Mentem is a safeguard of our freedom, for we are free by definition until proven guilty. The level at which an individual can be conscious of his acts and avoid trespass against another is the level of that individual's social maturity and his or her right to freedom. This maturity is a method with which to gauge an individual's social and psychic development within his or her identity. A mature individual will gravitate towards greater consciousness of his or her facts and create more of himself or herself in reality, to beautify both the self and the environment inhabited by that self. By contrast, an immature individual will gravitate towards coercion, to break or steal, to destroy that which is being created. Now, with the principle of Habeas Mentem, we can better identify which individuals within the group are, in effect, socially immature and thus tyrants, and which individuals are socially elevated in their conscious minds and thus are the builders of our society. Though there is still more, simply, we can now see the beginning of a law of consciousness in a still but semi-conscious world.

Chapter Eight

A Person in Agreement

A PERSON IN AGREEMENT is a real phenomenon. An agreement between two individuals is more than merely an agreement of their minds, their individual wills. Now, we can know that this agreement also encompasses their total personalities at infinity, their greater personal realities. A person in agreement is in agreement both with the mind of another and that other's greater identity. Thus, a person in agreement is the phenomenon of agreement between personal realities, between real identities.

In the mind, this agreement can be either conscious or unconscious. Conscious, it had been approached willfully, accepted, and thus chosen to be an act of agreement. Unconscious, the parties had been drawn to each other unwillfully, perhaps moved by curiosity of the other, or drawn to sensually, attracted to physical appearance or movement, or perhaps they had been brought together by what appear as chance circumstances. Such approach could be naturally, unconsciously in agreement until one or the other becomes cognizant of the approach; then there can result either a conscious acknowledgment of agreement or an expression of disagreement to stem further approach.

An agreement in a person courted unconsciously is approached naturally, much as an attraction in response to movements of the body, like a dance. The agreements form of themselves, spontaneously, and the attraction between the two bodies becomes more intense. One gives to the other, mutually, unthinkingly. The agreement may last only as long as the desire drives the two together and dissolve as soon as that desire is satisfied. Unconscious, agreement is almost beyond our control, fleeting, mercurial. Yet, it occurs, and can be very strong at times. It is exciting, thoughtless; it can be unreserved and even communicate in increased attraction without common words. It predates us as a specie and had exited before we developed a conscious mind. Agreement can have an almost predatory quality in its natural form, but it is an old value, it reaches down to the very drives that sustain our life, and that had been indispensable to our specie.

Conscious, we have the power to either accept or reject an approach. We are more complex now and our needs reflect that more intricate existence. Our agreements are more sophisticated and serve our needs in more subtle ways. What used to happen naturally can now occur also thinkingly. Where an agreement was an almost raw attraction between the natural needs of two beings, it can now have the added value of being an act conscious of itself. A mind conscious of its being is also conscious of when that being is in agreement with another. It can still respond to the same natural drives, but it need no longer be done to unconsciously; it is now a doer, aware of is actions, and thus is aware of when it is being done upon. A mind aware of its identity has a need that did not exist while still unconscious. It needs the right to consciously accept or reject an agreement; in effect, it needs the right to either agree or to disagree.

An unconscious mind is not as forced from its reality, its agreement, as is a conscious mind. The unconscious mind will tend generally to accept, provided there is some form of compatibility. It will generally yield when pressed or quickly weaken in its conviction and submit. The definition of its identity at infinity is not as distinct and can be easily modified to suit the new needs. As sincere and as beautiful in its naturalness and innocence as that mind may be, unless stubborn, it will tend to lack strong conviction and thus be more submissive. Its cry for freedom will be genuine but it would lack will and thus become quick to accept the trespass. Without a conscious mind, its identity still lacks a real definition at infinity to define it as a real personality. Rather, it is reflective of those personalities around it, easily modified, and quick to adapt itself to the dictates of a new authority. Its definition at infinity, still formative, can be easily influenced but it cannot be easily disturbed; if temporarily inconvenienced, it quickly retreats into its comfortable unconsciousness.

When fully conscious, however, we have a real need for our identity. To be forced from it, to be forced against our agreement is a real suffering that transcends our immediate discomfort. It is the pain of a soul that is being forced to be other than itself. A conscious mind is not slavish; it can yield, it forgives, but it cannot submit; its actions are deliberate, not forced, and it cannot bow from its right to itself, its right to seek agreement. A conscious mind aware of itself and in agreement with its identity does not seek to trespass; it is not tyrannical; rather it seeks mutually beneficial agreement. A real personality is advanced by successful agreements, but it is neither forced against nor forces another against his or her agreement. There is no advantage in disagreement since one party must exert effort to keep the other in submission; that same effort can be used far more effectively on other things when acts are voluntary and mutually beneficial. A conscious mind seeks to do things, no matter how difficult, in a voluntary manner because it chooses to do so; it does not do things because it is forced to. Forced against its agreement, a conscious mind holds firm in is convictions and rejects the trespass.

To seek agreement is a natural act. It is to seek those conditions of reality and in the reality of another that are compatible with the conditions that are part of our mind and reality, our identity. If we are in agreement with our identity, then we seek to preserve this agreement because it is a real demand by our reality as defined by the mind. As we mold the reality around us, our immediate environment and circumstances, we project our personality into it. How we handle things, how we possess them, and how we create new things are all indicative of the level of agreement between the self and its greater identity. But the materialization of the self need not be insulated from the identities and realities of others. We can be enhanced by things that belong to another, that are handled and part of another's reality. Then, there can be formed an agreement to exchange goods between the two personalities who find benefit in such agreement. The goods of each other's realities or services brought into each one's reality can then formally change hands, as agreed upon, with each to enrich the other. Both benefit from the exchange and the agreement is a successful one; both identities are then enriched with the presence of a new thing that is more compatible with their respective identities and thus are satisfied. We each seek to exchange something we desire less for something we desire more. Based on that is the act of agreement that occurs between personalities. The success of such agreement then becomes part of their now enhanced, greater personal identity realities. Each is enriched more by surrendering what is, to the self, less for what is more; but, if he or she wills it, each contributes of the self to the reality of another more than he or she receives. It is always a conscious act, entirely within the domain of how the self wishes to express its definition at infinity, for its own benefit. To be thus enriched is always an entirely personal act, to be judged by none than the self, and judged truly, really, only at that self's definition at infinity.

When we interact with other individuals, if they are conscious, we can depend on their act being also beneficial to them. An unconscious being may or may not be aware of the benefits of its acts; however, nor is that awareness, unconscious, of great importance, since its reality will accept more readily. By contrast, a conscious being depends on the act being beneficial if for no other reason than an unbeneficial act will detract from his or her reality and force the mind to turn its attention to that loss. It distracts the mind from other, more beneficial pursuits, requires attention and support until the error is corrected, and can potentially cause damage to not only one's own reality but also the reality of another. A condition of agreement is self supporting; it, through mutual advantages, insures its own perpetuation since it is in the interest of each to keep it so, and allows the mind greater freedom to pursue other needs. An agreement, properly executed, sustains itself since it is also endorsed by each other's reality. Thus, to a conscious mind, the element of agreement has an additional benefit of liberating it from the chore of having to continually preserve a state of conditions that are beneficial to the self. In agreement with another, a conscious mind, the act is itself self-supporting and self-perpetuating by the virtue that it is mutually beneficial to both parties and sustained by each other's realities. When consciously agreed, thus, neither party trespasses against the other and, on the contrary, the identity and reality of each is mutually enhanced.

A conscious mind, as opposed to an unconscious mind, is a mind that is continually conscious of its responsibility. It is responsible in its behavior both to itself and to another. In agreement, it seeks like mindedness, since it needs confidence in the responsibility of the other. Then, agreement can be formed and both can benefit from the exchange; if no confidence of like mindedness is found, no beneficial exchange can take place. If entered irresponsibly, in error, the agreement with a still immature mind can run the risk of disagreement and possible trespass. A still unconscious mind can be a liability detrimental to one's reality if sought on equal terms. Given that mind's still formative identity, it can reverse its posture through a loss of conviction, it can be easily swayed in its thought, and even possibly given to dishonesty and steal for a quick gain. Then, the mind thus trespassed is thrown into a disadvantage, is betrayed in its agreement, and must seek its reparation to recover its position within its reality, to reestablish its presence in its identity. An unconscious agreement, if at times appealing, can have a serious disadvantage to a conscious mind. An unconscious agreement can be irresponsible and dangerous. It is a danger for which a conscious mind must be responsible and watch for.

Thus we see that there is a principle to agreement. It is that a mind is free to be itself, free to seek agreement as it wishes, as it consciously will, provided that it is conscious of its acts. A mind is free to occupy, or seek to occupy, its identity provided it does not seek to trespass on the identity of another. As long as we do not force another against his or her will, as long as we are not the author of such force, then we are free to seek voluntary association as we please, and as it is agreed upon by the parties who enter such association. Provided they are in agreement, then all concerned are better occupying the value in reality that is their identity. To occupy this identity in agreement is a natural act of our conscious mind. It is a conscious act to recognize the value of another's mind and his or her conscious identity. Conscious of this act, not forcing another against their will, not forcing a mind from its conscious identity, seeking the benefit of agreement over the destructiveness of disagreement, the mind is conscious of the principle of agreement. Conscious of this principle, acting out this consciousness in its interaction with other human beings, interacting willfully only through agreement, it is itself by definition; it is free.

There is no social tribunal that can define whether or not a human being is free. We are all free by definition until we break this definition and step from freedom by breaking the principle of agreement. The principle of agreement is a law all unto itself. We are defined as human beings, conscious of our acts and conscious of the human value of others as long as we obey the principle of agreement. As long as we do not trespass on another, do not force another against his or her agreement, then we are free as minds within our identity, as human beings conscious of the Law of Agreement. To break this law and to force another against his or her agreement, is to be unconscious of the value of the freedom of each individual mind, of each person's identity in the universe, and thus to be less than fully human. Not human, we are not free to be ourselves in a human society, and then we are subject to those laws that seek to restrict our freedom of action within society. We are free only while we are conscious of our acts, while we obey the Law of Agreement, while we are in our mind. Unfree, we are not a definition at infinity unto ourselves, are amiss with ourselves and others around us. Within human reality, we are free by definition until our acts prove us unfree; then, we must stand before a social tribunal and seek to regain our freedom which will be granted only after we have paid our dues to society and the persons we have injured.

A man guilty of trespass is not unfree until tried by social law and found guilty of trespass against another. An accusation does not condemn; a man so accused, if he or she be a conscious mind, will stand defense, in his or her mind, until either proven guilty or remaining innocent. A mind conscious does not submit to tyranny. It cannot be accused lightly and be expected to surrender. It is not slavish. A mind conscious of its acts and believing itself innocent will persevere if trespassed upon by false accusation until it is proven innocent. In this lies the strength of the Law of Agreement within Habeas Mentem.

Freedom is a responsibility of the conscious mind. A free society can exist only if the individuals within it are conscious of their identity. It cannot be formed for the benefit of a servile population, for they would be unconscious and could not defend themselves from the advances of tyranny. Freedom exists only when each individual is conscious of his or her mind and of his or her value as a human being; then they have the strength with which to defend this value when approached by coercion. In that is the strength of a free society. It is composed of individuals who are free because they allow other individuals to be free themselves, because they interact with each other only when there is a bridge between them of mutual agreement. But where this bridge of agreement does not exist, they do not advance and force themselves on another until such agreement is reached.

A conscious mind moves forcefully, boldly, but it does not trespass and is not trespassed against. In that, it is responsible as a social mind, because it is responsible first to its greater identity in its universe.

The definition from the universe that communicates to that mind and its reality is a definition from a universe that is working compatibly with its identity. This compatibility between the mind and its universe becomes physically apparent in both the structure and disposition, uprightedness, of the mind and the elevation of the reality that is its environment. The materialization of this definition socially, within its immediate reality, is a real definition of circumstances and forces that are now working with it as it moves in life. To be in agreement within infinity is to be one with one's identity at infinity; it is also to be in agreement with a universe that now has the bridge, the medium, with which to materialize its greater definitions within our life. An agreement, when formed between individuals, is a materialization of universal order in the realities of their lives and the realization of our mind in our social reality. Our physical environment moves with us in our agreements and our society reflects more clearly that value that is our human identity. We are the channels, through our minds, between the universal order and the order of things in our society here on our planet. We can open these channels only when we are in agreement both with ourselves and with each other. To seek these agreements genuinely, consciously, responsibly, forcefully, is to materialize the value of our human mind, our definition, in the even if otherwise inert reality of our planet. In the Law of Agreement, in Habeas Mentem, is the bridge from our past into our more conscious, more living, future in our reality; in them is the social law that has embodied in itself the universal law of its own, self defined order as it can be translated into our human reality.

A man or woman in agreement is in effect a person in tune with the principle of his or her greater reality. To find agreement where such exists, is to materialize in the self and in that self's environment the image of that self's greater identity, within and without, at infinity. To have the mind is to be in infinity as infinity is that mind and its Earthly reality. Can the mind know when it is itself in infinity? It can be itself when it is conscious and free from trespass. Can it be conscious and free from trespass? It can be so only if it obeys the Law of Agreement and is protected in its mind by Habeas Mentem. Can the Law of Agreement and Habeas Mentem be a social reality? They can work in society only when we are conscious of them and know how to apply them to our social order. Then, our social order reflects the principle of man in tune with his or her greater identity and then the universe reflects its infinite order in the materialization of our Earthly society. Can we know what that materialization is when it becomes our reality? We will know it when our planet becomes richer, more beautiful, more in our new image. Then, in our human being, reality will materialize in us the answer to the question: "We are Who?"

Chapter Nine

How Do We Build a Society?

HOW DO WE BUILD a society based on the principle
of universal order? If the Habeas Mentem is the definition
of the individual as a free mind in the universe, and if the Law
of Agreement is the application of this definition in our social
reality; then, how do we build a society based on the principle
of Habeas Mentem and its corresponding Law of Agreement?

First, let us examine some existing conditions that will affect
the method of our construction. Then, let us examine the principles with which we may work and allow us to act in ways
consistent with our definition as a free mind in the universe.
Based on these, we can then judge which recommendations
for change will be in accordance with these principles and which
changes will be contrary to our human being. From these we
can draw conclusions and examine which structure of society
will best suit the needs of a novel, emerging conscious mind
of man.

One condition of which we must remain conscious is that
already existing societies are natural phenomena of our planet.
These societies, whether they be European or Asian or African
or American, and whether they be more tribal or mercantile
or whether they are individualistic or communal, are all indicative of the social mind as it has materialized in that society
through time. Each society is of necessity indigenous to its people and its environment. In response to the existing reality and
to how the mind perceives its reality, how it understands its
universe, the society formed around that mind will reflect the
collective perception of reality. Each society will exhibit traits
peculiar to its people and their immediate environment. If they

are benevolent or cruel, wealthy or poor, liberal or regimented, these are all characteristics that have become manifest around that people in response to how the mind had accepted itself as a social reality. In each instance of life, the mind either chooses to respond and act, it chooses willfully, or it chooses not to respond, not accept responsibility and instead be acted upon. Through these conscious or unconscious choices the mind materializes its environment in proportion to its judgments in reality. The personality of a society is then a reflection of its collective mind and the persons who had learned to coexist within that mind have, in effect, come to terms and agreement with the limits and benefits of their society. The mind has accepted, in effect, submitted, to the resulting social order. To not do so would either put it at odds with its society, in which case it would either have to act to change it or to withdraw from it; or it could agree to coexist within its reality. To do otherwise would, historically, cause it to die. As long as society reflects, predominantly, its consenting and surviving individuals, then it must reflect the personality of its people.

A society in agreement with its personality is in agreement with itself. Provided there is no outside influence, such as conquest or subversiveness, then the behavior of its citizens would be a self centered activity. Each person responds to his or her circumstances within the conditions that govern their community such to benefit them individually. These responses are acted out according to their mind in a way that results in the social character of their individual world attuned increasingly to how they are. As each person contributes to and is contributed to within his community, he or she gains a stronger sense of unity within that society and the community takes on definable traits that are peculiar to that people. In reality, the social order becomes more as are its people unified into an increasingly greater sense of community in agreement with itself. Thus, to an outside observer of this self agreement, whether or not it meets with our approval, because it is a real definition of that social reality in relation to the definitions of the mind of its people, we must accept it prima facie. That society is as it should be.

Therefore, our first principle must be that we do not force change in society; entering it as an outsider and seeking to improve its sense of order with the universe, because that society is a real representation of its people, that society is special and a real definition from reality; we may not change it. What transpires in each social order may or may not agree with our principles, but, unless invited to do so, we may not enter it to change it to suit our sense of correct order. We may have the right to communicate with it, to trade, to introduce our ideas, but we may not change it from within; that is the responsibility of the people native to it.

Thus, to build a social order based on Habeas Mentem is not to change the world. It is not a missionary task nor one of conquest. Change, to be such that would be consistent with our philosophy of order, must be indigenous and desired from within. The purpose of this philosophy is to offer an alternative and to offer direction and solidarity of thought to the mind ready to reach for it; it is not to be imposed on anyone against one's will. To do otherwise would be more damaging than helpful. It would be destructive to the delicate balances that had already established themselves within the social order, which would precipitate economic chaos and political instability; also, it would destroy the representation of reality that had formed itself in relation to the minds that are that society. Change can be brought about without the need for revolution, without destroying the old order to rebuild anew from the rubble; that is a very wasteful method and one which can require generations for society to once again reestablish itself in its reality. Forceful change is harmful to social change and destroys the delicate balance between a people's collective mind and its established social reality. But there are other dangers.

If revolution is insensitive because the change is so abrupt, change advocated over time by a manifesto can be equally harmful to a people. A manifesto is an artificial demand on reality because it demands that man suit the designed social order rather than to design the social order to suit man. If a decree defines that man is or should be a certain way, then that definition is man made, artificial, and to demand that the mind of man adjust itself to that artificial definition can be damaging to that mind's identity. We do not know what is the definition of man, since each individual has a separate and unique identity that is defined only to the self by that self's definition in reality. Under Habeas Mentem, it is as defined by that self's interrelationship-definition in the universe. Thus, man's natural definition of his or her identity is first, entirely personal, second, entirely as defined by the universe. Without a general definition, we must then work within that context and accept our individuality and universality and design our social system accordingly. A person, provided he or she does not trespass on the reality of another, is defined entirely by his or her reality. That is the difference beetween the concept of Habeas Mentem and one of a social manifesto: In a dictated program for social change, a person must be trespassed against for the benefit of the prejudged beneficial social order; change according to the Habeas Mentem would be where society changes to not trespass against the individual. Under the Habeas Mentem, the change of society would be to suit the individual not as defined by man, since an identity is beyond general human definition, but as defined by the universe. That definition is apparent in the materialization of the human reality. Then, the purpose of change is to make that reality and its apparent materialization more pronounced.

Now we can have direction for social change. We do not advocate change to societies how they should become. Rather, we would advocate change to have societies become more as they are, in effect, to magnify their social definition and as it affects each individual within it. Then, with the aid of this better self observation, each society can identify where lies its weaknesses and its strengths in relation to itself as it wishes to be. The purpose of this change is to create a more realistic materialization of existing conditions from which can be gained social data that would be meaningful to a social scientist. Same as a scientist seeks reliable data from which to draw conclusions, so does a society need to expose those facets of itself that camouflage its activities in relation to the activities of the individuals who live and act within it. Then, human action becomes more evident and the mind can judge whether what it does in reality is materializing in its environment those results that are desirable to it.

The question that follows naturally is that, if the various social masks that camouflage the real consequences of human action in society are removed, can society function? If so, can it function well, or at least better than if these masks were left untouched? To answer these we must, first, know what these masks are and, second, we must know what it is that makes a society function.

These social masks exist wherever there is a restriction on human agreements that are beneficial to the individuals involved. These may be controls that restrict the flow of information, rules that inhibit the formation of agreements, and those that force individuals to break the Law of Agreement. When individuals are unable to easily form agreements between themselves, agreements that are non-coercive to third parties, then the advantages of mutually beneficial exchange between individuals cannot as easily be realized. When individuals are unble to form agreements between themselves, not protected from being forced against agreement into disagreement, then society is handicapped by being unable to materialize in itself what it is that individuals desire most, which is most their identity, and that which is most advantageous to them. Then, individuals are forced to act according to how someone else wishes them to be rather than how they would wish to act themselves. They become done to rather than being encouraged to do, consciously. Done to, society then becomes other than itself. To a conscious mind, to be done to becomes an unbearable state of affairs that gradually degenerates into covert activity. Ultimately, it would regress to social activity that seeks to negate the results that had been judged beneficial by those who had imposed the various rules and controls. With constrictions on agreement, society is forced to become other than itself and it becomes difficult to judge realistically whether what we do is what we materialize in our reality. We are as we do, unless we are done to. As conscious minds, it is our responsibility to do and not be done to. Thus, as conscious minds, it is our responsibility to identify those masks that prevent human agreement from materializing naturally and advantageously.

The question remains: Will society so unmasked, stripped of those inhibitions to human agreement, work as well if not better than the present state of affairs? The answer lies in what it is the citizens of that society desire. If the social masks, in effect, social myths, afford them a sense of comfort, then they need not abandon them. If these myths are recognized as such and are accepted, willed, then they are the image that is materializing in response to the social mind. However, it is also then the responsibility of the conscious minds of that society to recognize that these myths do exist and that they tend to distort the resulting reality. If, for example, the general public wishes to be granted a certain favor by the government that would lower their cost for some particular service, they must be prepared to face the eventual consequence of either the diminution of that service or of their payment for it through rising costs in some other quarter, possibly raised to some portion of the population that gets no benefit from this favor. A conscious mind can see this easily, whereas an unconscious one may not; but it is the responsibility of the conscious mind to know these things and understand how these myths may distort the social reality. Until this myth is unmasked, society as a whole may have difficult understanding why a certain action on their part would yield results entirely contrary to expectations. Unmasked, then the public can better decide whether it wishes to continue in its myths or discard them. Then, whether society will work better will be based on that decision.

A society unmasked will materialize what the social mind is really like. It then has no facade behind which to hide and with which to take little advantages from society at others' expense, because no agreement had been established which would entitle a person to that advantage. A society based on reality is unforgiving in how it materializes our actions around us. As we do, so is it done to us by reality; as we form our agreements with our fellow man, so we must live within the materializations of those agreements; if we are insincere in these agreements, then what will materialize will be other than what had been expected. If we seek to hide behind myths, then we must accept that we will have to confront the consequences of our actions indirectly, probably unexpectedly, and ultimately unpleasantly. In a real society, the mind must be mature and conscious; it is the mark of a free mind. A free mind must be aware of its actions and their consequences, responsibly and sincerely; irresponsibly, it is then forced to suffer the consequences of its action and ultimately lose more than it had gained. However, if it is our choice to be irresponsible in our actions, insincere, then it must be our choice, by default, that we would rather materialize in our environment immediate benefits at the expense of more permanent ones. If this is then recognized as a myth and so accepted, then it ceases to be a myth because it had been unmasked and becomes a conscious, free choice. What materializes from such choice is then fully accountable and of surprise to no one.

So now, having established some background conditions, we can seek an answer to the question: Which society will work better, masked or unmasked? The answer lies in whether or not that society is ready to seek change. In general terms, the specifics of which we will examine later, we can now see that some societies would perhaps work better unchanged, whereas others would be enhanced by seeking to make their society more directly responsive to their human action. Not all societies would be comfortable in the stark nakedness of real materialization of their actions. Sometimes, it is beneficial to experience things indirectly, shrouded by some fantasy, social comforts that would rather be paid for at a future time, in a different way. The conscious mind understands these things and acts accordingly; the unconscious mind must yet learn their meaning. Whether a society elects to materialize its reality directly or indirectly is a decision left to its social mind. If it is more comfortable with its myths, then it must accept as part of its reality the consequential handicaps; change cannot be forced upon it. Ultimately, as the need for the support of social myths dissolve, that society will also change to make itself more directly responsive to its actions. Then, it becomes the responsibility of its actions to materialize the desired results. Change cannot be forced; it must come from within. If, however, a society remains unchanged, it is the responsibility of a conscious mind to understand the meaning and disposition of that society and to act with it accordingly. Same as we would not impose social change on another by manifesto, nor would we allow that society to dictate to a society already free.

Chapter Ten

Principle of Habeas Mentem

A SOCIETY BUILT on the principle of Habeas Mentem is a society based on a principle that is consistent with the way we view our universe. Our universe is now no longer a manifestation of random forces and accidental circumstances; now, our universe is a natural order formed for its own purpose and moved by its own knowledge. We can harness this order to have it become the foundation of our new social order by consciously seeking to occupy the definition of our identity; we can occupy more closely this real definition by applying the definition of our human identity to our human organization through the principle of the Law of Agreement. In agreement, we materialize the natural order of our universe in the midst of our human definition of our reality, our social reality. Thus, a society built on the principle of Habeas Mentem is a society built on a principle of human agreement. Let us now examine the nature of these agreements.

A person in agreement with another is two; a third is needed to form a society. Each person in agreement with another forms a social bond that is beneficial to each party involved in this agreement. In this manner, agreements between individuals would never cause social friction, if they never had the power to affect a third that is not part of that agreement. The fact that an agreement between two persons can affect a third is the basis on which are founded nearly all laws that regulate human behavior. It is the ability of an agreement to trespass on others that can limit our right to freedom, which can then limit our right to seek agreement. It is that simple and, yet, it can have the power to control entirely the formation of a society.

A law is an agreement formed to keep another in check. Whether the law is agreed upon by the king and his ministers or by a majority of a legislative body and endorsed by the people, it is nevertheless an agreement that will define future action. Future human action will now be contained within the limits of that law. If that law seeks to limit trespass by enforcing an agreement, then it is beneficial to those who wish to enter such agreement. If it seeks to shelter those who are not part of an agreement from encroachment by that agreement, then it protects individuals from trespass by a group, whether it be only two or more, and thus protects from the potential tyranny of the group. In both cases, the law seeks to prevent trespass and enforce agreement. However, if the law seeks to trespass against an individual because of an agreement by a group, then we have the kind of law that is necessitated by the formation of three, a society.

In a society it is necessary to pass laws that define behavior for all individuals in relationship to the behavior of the group. All individuals, for example, must pay taxes for the benefit of the whole. None are exempt, except by special agreements, and in almost all cases the payment of taxes is an involuntary act. We do not have the luxury to finance the functions of government out of voluntary contributions. The agreement that defines our government, namely the constitution of the country in which we reside, defines that each one of us will be assessed for a tax. We are generally not given any options as to whether we must pay this tax, unless perhaps if we support through contributions some other social functions, that we be exempted from paying the full tax as assessed. But, again, this is as defined by the social agreement. Thus, unless the social agreement makes provision for a variety of options to which we may contribute freely, of our own accord and as we so will, in lieu of a tax paid directly to the government, we must pay as it is levied by the social agreement against us. Tax is a universal example of an agreement that, through its definition, has the power to force another against agreement.

Not all individuals are favorably disposed to the payment of taxes; probably few are. But few would argue that taxes are unnecessary, since without them the government would fall.

Unless one is disposed towards anarchy, a society without governnment is not a desirable state of affairs. Thus, though a person would pay taxes grudgingly, he or she would nevertheless prefer a state of things as is made possible through those taxes rather than a state where no social order exists. In effect, if only by default, a person pays taxes because, though he or she may feel that they are being forced against their agreement, if they wish to pay less taxes or have that money used elsewhere, they nevertheless agree to pay them as levied because the alternative act would be less desirable to them. The alternative might be either fines or imprisonment or, if their government is particularly liberal towards non-payment, the collapse of their social order through a lack of financial support.

Now, an alternative might be that an individual, either at birth or upon maturity, is either declared or declares himself incapable of abiding by the agreement thus forced on him by society. At one extreme, that individual would thus not be free to partake in any benefits of existing in such a society, without protection of any of its laws and services; or at the other extreme, the individual becomes a ward of the state and is entirely at its mercy. If the first option is exercised, the person has in effect, relegated himself to a wilderness where he or she must arm themselves against predation, must resort to a kind of barter economy because there are neither guarantees of contract nor of money, and can own no property other than that protected by their armaments because there would be no formal guarantee of that ownership. In reality, such a person would probably not be allowed to live in a civilized society, though some civilized societies can at times approach such a state of affairs, as in war or total deterioration of the social fabric, and the individual would have to seek existence in some form of exile, such as in the wilderness. This way of life may be appealing to a romantic nature but few individuals would opt for it as a permanent alternative to civilization. If the other option is exercised, then that person also has no rights because he or she is entirely the ward of the state. Their freedom is now defined by how the state wishes to define the limits of their activity and their inter-human agreements are restricted because, as wards and as having renounced the right to agreement, their agreements with others are non-binding. They are forever, in effect, prisoners within their societies and, even if favorably taken care of, have no freedom other than that allowed them by their social masters. Though both alternatives can be approached somewhat within the established social order, if there is still wilderness or if there is a benevolent welfare state, neither alternative is desirable to the greater majority of the public for any length of time. Perhaps for a small fringe group these alternatives should be encouraged to exist within the social structure where a man can escape temporarily and withdraw from the social agreement, as a hermit or perhaps in exchange for a lump sum payment he can be taken care of for a period of time as a total ward of the state; but these conditions are not the normalcy unless society has become such that it drives an increasing number of individuals into these extremes of non-agreement. But they are the exception under normal conditions. When faced with the option of whether to agree or to not agree with the general agreement that is society, as in the case of taxes, we generally choose to agree.

Thus, society is an agreement, and it is a voluntary agreement not only of its majority but, more likely, of its statistical majority where the greater portion of the population, tending towards normalcy, tend towards a social agreement that is generally acceptable to them. Based on that, then, the decision whether to pay or not to pay taxes is to be evaluated as an agreement of a majority, if it is a democratic government that levies such taxes with the public's consent, or of a statistical majority because, by default, the tax is paid. If it is not paid, then the statistical majority has opposed it and the agreement that is the government is in jeopardy. The normalcy of human agreements is broken because the consequences of non-payment, if they be severe enough, have made such resistance unattractive and which would occur only if the conditions that describe the agreement are truly odious to the public at large. Such is the general agreement that is society.

A society is an agreement of its majority. In a democracy, it is formally the agreement of its majority as shown by ballot; in any society, it is the agreement by a statistical majority of the population by consent to is laws. If a society is experiencing disagreement by individuals with its laws to the point where that level approaches a statistical majority, it either will find it difficult to function or it should reform. This may sometimes happen for a small group within society at large; in that case, that group should either be allowed to leave or to reform itself under a separate agreement either with or without the support of society at large. But, more importantly here, the social agreement that allows a society to function favorably is one that is endorsed by the actions of the statistical majority of its population.

Therefore, what is the permissible level of trespass by a group against an individual? Because the group is a voluntary association formed for its own benefit and defined by that group's laws, the individual must accept the agreements as they are defined by that law provided they do not consciously force the individual from pursuing his or her identity as dictated by their conscience. In other words, should the resulting trespass of the group against the individual be such that it forces the individual to willfully break the law in the face of that law's punishment for disobedience, then that individual is taking a conscious act of opposition to the social agreement. Such opposition may not be taken lightly, since it is a serious act that is potentially harmful to the agreement that is society. In some small measure, it can be argued that to disagree with the majority is to force the social order into disagreement, which is in itself a form of trespass. For this reason, a fickle disobedience of a law should be punishable, since it is not an act conscious of its consequences. What we are examining here, when we ask what is the permissible level of trespass of the group against the individual, is a situation that requires conscious attention by both the group and the individual.

Thus, when there is a conscious opposition by an individual to the agreement formed by the group, that opposition must be examined directly and put to the test of law. The individual must prove to the social court that the group's agreement is a trespass against his or her identity; the burden of proof of trespass, in the case of social agreements that are written formally into law, then rests not with society, but with the individual who claims to be so trespassed. If the individual's refusal to comply is in fact found to be fickle, mindless and without conviction, then it is simply a matter of social disagreement, disobedience, to be punished as provided for by law. If, however, the social law is in fact a genuine trespass to which the individual is consciously and painfully aware and in opposition to, then it is the burden of that conscious mind to prove that the law is a disagreement with its identity, that it forces it into disagreement with its mind, against that person's conscience, and that he or she be either exempted from it or that the law be modified such to restore their freedom and allay their grievance. Under Habeas Mentem, this is a condition that is forever present, because in each law written by the social order there is the potential for error and social tyranny. As a guard against this potential tyranny, the free mind of man must have the right to challenge any law that defines its society. If a sufficient number of individuals, so challenging and putting to the test in a court of law, are in fact found to be trespassed against, then it is the responsibility of free, conscious minds to recognize this trespass and, since it would benefit society, to change that law such as to cause less opposition. If for no other reason, it is sensible to do this, for to do otherwise causes further disagreement within the social order to such a degree as to be damaging to the social fabric as a whole. Thus, these are the limits of trespass by a group against an individual.

This principle of how social law works, where two can force a third, can apply to the social agreeement only; it cannot also apply to how individuals may act within this in their dealings with one another. By law, individuals have the power to form their own institutions that have power over their members, such as clubs, companies, or other private societies, but they do not have the power to coerce, which is a power relegated entirely to government. Private associations are formed by agreement and the individuals who wish to belong to these associations must in some manner obey these agreements and even be penalized for not obeying them, if such punishment is part of that association's agreement; but if a member of a club or company or other private society decides that he or she no longer wish to accept and abide by the agreements that define their membership, then they have the option of leaving that association. This is something that generally we cannot do in a society with regards to our laws; we must obey them, except when we are individually exempted. Thus, individuals do not have the power to coerce. All agreements are voluntary unless they are written into enforcement by law as a contract. Then, to break the terms of a contract so written is an act of social disobedience requiring reprisal not by the hand of the individual so wronged but by law. A contract is an agreement stated such that for either party to fail to fulfill the terms of that contract represents a trespass against the other or, in effect, a disagreement. In that respect, though individuals do not have the power to enforce their agreements, except for where in the absence of government they are forced to take the law into their own hands, as in combat, they do have the power to entrust their agreement, if in the form of an acceptable contract, in the hands of social law and its power of enforcement. Individuals are free to form agreements between themselves but, except under certain special circumstances, as in self defense, they are never free to form agreements that seek to trespass another and force that other into disagreement. Thus, except where there is no existing social order such as in the wilderness, individuals do not have the power to form societies unless they be private associations that are totally non-coercive to its non-members. In effect, individuals are allowed to form associations only for their benefit, as that benefit may affect each member and as that member is willing to voluntarily submit to that association's by-laws, but they can never form an association for the benefit of a third party against that party's agreement. That function is entirely the domain of government.

Now we can distinguish between agreements between individuals and their government and between individuals and themselves. Only governments are allowed to form agreements that have the power to coerce another. This coercion can be permitted to the level that is permitted either by its constitution, as endorsed by its formal majority at ballot, or as agreed upon by the statistical majority of its population, by default of obedience. However, it has no justification over the life of an individual if that individual is consciously trespassed, forced into disagreement both with himself and with others, by that coercion. Then, however, since society had been organized for the benefit of the individuals who are in agreement with it and for the purpose of insuring each individual from becoming forced against his or her agreement, against trespass, the burden of proof rests not with society but with the individual so trespassed. On the other hand, agreements between individuals, except as defined by contract, may not have the power to coerce each other; they never have the power to form agreements that are designed to coerce another. Such is the agreement that is the social contract, otherwise it is not a society by agreement.

Chapter Eleven

A Society of Individuals

A SOCIETY OF INDIVIDUALS is a social agreement most consistent with our definition of man as defined by the universe, by his or her identity. It is a social order formed because individuals form agreements between each other and collectively amongst themselves. These agreements then formed, those that express our identity and are protected from coercion, are those that allow us to express ourselves as we really are. In our communications with one another, our exchanges of goods and services, our gifts to one another all reflect a mind free to be itself in its universal order. Such a society of free individuals is that social order that is most consistent with the free mind of man.

Individuals are free to exchange by agreement and, by definition, are free to choose how they will exchange. They are free to choose those agreements most beneficial to themselves and to respond to their choices as these materialize in the consequences of their future. When free, they are able to choose those things that are most like them and avoid those that are most alien to their personality. If they choose wisely, consciously, then their choices will have beneficial consequences for a long period of time; if they are careless or foolish, what they manifest will be small and short lived. Thus, as they choose, they materialize their personality in their society. If it is great, mature, beautiful, then it is the mark of a great and mature people; if it is lowly, degenerate, ugly, then it is a reflection of a people not yet having reached maturity. To become mature as a society is the mark of a people with the strength to face reality and choose consciously the next step of their human development.

In our past, when individuals first learned to exchange rather than to take by force, the world changed again. They learned the value of an agreement over the impulse for a quick gain by force. In that first exchange was formed the foundation of conscious human acts that are beneficial through time, of a society more suited to the more conscious and more modern mind of man. In that first exchange was the first foundation of a society of individuals based on the principle of agreement.

To take by force, to steal, is an unconscious act. It is a primitive impulse of acquisition that, as matures the mind, is sought to be controlled and finally overcome in a civilized society. Society cannot function in a progressive manner suitable to an advanced man if the value of each individual is trampled by the right to steal and take by force from those individuals who are most productive and most creative. To take by force is negative and coercive and can succeed only in a servile population tyrannized by an order of unconscious minds. If it had been our past, it need no longer be our future. As conscious minds we must be conscious of the consequences, through time, of our actions.

To exchange is a conscious act that does not seek to trespass on another against his or her agreement. In each such exchange, we become conscious of the future consequences of our acts and strive for beneficial results that will open the door for future agreements. It does not exploit and run away like a thief with his prize. To exchange in agreement is a recognition of the value of each individual and that individual's right to property. It is an act of surrender, of giving, where one surrenders what one values less in exchange for gaining that which one values more. As a mutual act, both benefit, each giving what is theirs voluntarily to the other.

An exchange goes beyond the act of giving by agreement. The goods so given transcend the immediacy of the act. They are goods that are altered in their definition in reality by their possession. They are owned, handled by a personality, and become altered again when they are transferred physically into the possession of another by agreement. An agreement, we can now know, is more than a confirmation of compatible minds; it is also an agreement of their respective greater realities.

Our possessions are a part of our personal greater reality. Thus, the goods exchanged, being part of each owner's greater reality, when given over to another voluntarily, is to transfer the owner-ship from our reality to the reality of the recipient; in exchange, we gain a portion of the other's reality. The compatibility of realities is thus expressed at two levels of our reality: as ex-pressed by the conscious agreement of the mind and as defined by the greater reality that is the identity of that mind. Exchange, if simple, is an agreement of minds at one level, but also a much greater agreement of the dimensions that are our consciousness out there at another level. If an act of exchange appears effortless, almost automatic, then it is only because the act is so natural to us; at the limit, at infinity, it is an act that is almost infinitely complex.

Thus, through Habeas Mentem and its Law of Agreement, we are able to condense, in principle, the initial concept of an interrelationship of space and its corresponding universal defini-tion of a human identity into a real definition of man, having a mind, as a free conscious being; the resulting concept of these universal definitions through the principle of interrelationship can then be further condensed into that principle that allows each mind to be more itself and create its identity in, and become more, its reality; it then becomes, in its final analysis, the principle of exchange by agreement.

To exchange by agreement is a simple, almost mechanical, act that enables the universe to materialize its value in our midst as its definition of man as a creature that exists and creates in its image. It is that simple; and yet, it is that complex. When we act simply, sincerely and innocently, when we seek to bet-ter or to please rather than to hurt and break, to seek to find agreement rather than twist by force, when we exchange or give rather than steal; they are all representations of our human-ness in a way that is infinitely complex. They are an expression of that value in the human mind that is the principle of the order of our universe.

Thus, there lies our future development. We are traders rather than thieves or conquerors. We do not harm consciously and withdraw if we have harmed unconsciously. We demand, as conscious minds, that others do the same both towards themselves and towards others. In a society of conscious human beings, of individuals, these are natural laws that help us agree and protect us from forced disagreement. When these principles are recognized and respected, then our social environment becomes safe for the conscious mind and the social order functions smoothy and progressively on a principle of exchange and agreement.

As conscious minds, we are possessive by nature. What we own is ours, consciously ours. What we fashion from raw matter, how we create, what we gather or grow, are all a part of our mind how it fashions itself in reality. Our possessions, provided they were not torn from another by force, are all creations of our identity's self expression. What we gain through trade, through agreed upon exchange, is an expression of an acquisition gained by agreement. What we make, if we do not surrender it in exchange for something other, is part of that which surrounds us. What we hold, own, what we cherish in our reality, is our possession of it because reality has entrusted itself, there, to us. It is our real property to be claimed by none other than the owner and to be separated from by none other than agreement. We have a natural right to own that which is ours.

Conscious, we are more complex. We cannot surrender by force. To be forced to surrender that which is ours, what we cherish and care for, is to force pain that transcends the present. We adorn our reality with what is entrusted to us, what we handle and create in our image. To have it taken by force is to sheer the bond that connects us to our greater reality; it sheers that which we love. We cannot abandon them mindlessly, for they are our responsibility; they are part of that value that infinity has ascribed to us. We are physical beings in a physical universe condensed in our mind as the value of infinity that is the identity of the spirit, of our being. How we possess things is how we lend that spirit into the things that we possess.

But not all things possessed are so intimately ours. We do not possess all things at the limit, so deeply. Assets, property, can be held for trade rather than for intrinsic possession. We do not possess always for the sake of possession; some possessions are held in trust until they be exchanged for other, more desirable possessions. These are decisions that can be determined only by the person in whose trust they are held; none other can make that value judgement for them. Regardless of how a property or asset or right to such property is held, it is always left to the judgement of the individual in whose possession it is with regards to its value as a possession. Some property may be held only for the purpose of exchange because, if for no other reason, one cannot exchange with that which one does not own. Exchange can usually be realized best if the property so exchanged is not tied to a strong emotional possessiveness. If that is the case, then the price of exchange may be very high and transfer of such goods may occur only when that price is met. Then, what is being exchanged has a strong personal value and can be surrendered only at great sacrifice. However, in most cases of exchange of an economic type, the subjective attachment to the assets being exchanged is minimal, oftentimes to the point of being negligible. Then, exchange can be a simple and beneficial social act with economic value.

In a society of individuals based on a right to property and the right to exchange, the act of exchange is a mechanical social process that works because it is natural to us. The more sophisticated the exchange, the simpler the mechanism of exchange that effects an impersonal transfer of goods and assessment of value. What results is an economic system of market exchanges where trade is taken from the level of interpersonal barter, where the likes and dislikes of personalities may affect the outcome of exchange, to an efficient and impersonal market place. Market exchanges are where those personal decisions that each individual decides in response to those conditions that face him or her at each moment of time are translated into the decision of whether to buy or to sell. What we need or what we are able to offer are translated into the action we take in response to the reality from which we must choose our actions; on an exchange where such choices are translated into economic activities, such decisions are carried out in an impersonal setting of activity where personal recognition plays virtually no part, and is actually unnecessary to its success. Markets do not work because people know one another; rather they work best when they do not know one another and are able to carry out in a businesslike and impassionate manner the needs of exchange. A pure market system works without biases and personal prejudices; it is a forum of buyers and sellers basing their decisions on economic value and circumstances. It is exchange and agreement brought to its simplest principle. It works because it is a most direct and empirical expression of how individuals succeed in expressing themselves in relation to their personal economic conditions. Market exchange is simple and unbiased, if protected from coercion, and is the closest empirical expression we can find that can be made to measure human action.

There are many levels of exchange, and each individual so participating should be allowed to seek that level of agreement that is most beneficial to his or her needs. If the exchange is a complex agreement that will tie personalities together over a prolonged period of time, such as a mutual business venture, then it would benefit the parties involved to at least have some personal compatibility. It then becomes as much an exchange of personalities as of skills and assets and should function at that level of agreement. If the exchange will be one that is relatively simple, as that of an employee of a company or organization who exchanges his or her labor and skills for a wage or some other renumeration, then the agreement that exists between employer and employee is of a more simple fashion and one that should be less influenced by personality. It is an exchange of skills and, in return for these skills and productive labors, of wages, bonuses, advancements, and so on. Other than the personal compatibility of fellow employees, the role of personality should not be a major influence of the agreement that exists between them. Exchange should always be kept simple, if possible, and without strong emotional attachments; it benefits us most when it is least personal and most businesslike. In the market place, to gain the best advantage of the economic contributions of the principle of exchange and agreement, the mechanism of exchange should be kept at its simplest; then it approximates most closely those economic values between individuals in relation to their contribution to the socio-economic whole. With their economic activities expressed as agreement in a market exchange, the resulting values empirically expressed, through their medium of exchange, a price then expresses through a common denominator their economic needs and the value of their productive activity.

A system that allows us the freedom to trade and to find agreement amongst ourselves in relation to our needs and to our ability naturally materializes into a market system. That system, when brought to its greater conclusion as an impersonal mechanism of exchange, and when protected from potential coercion of its participants, will materialize as an economic mechanism that easily and accurately transfers goods and services from where they are needed less to where they are needed more. It works because each individual has the freedom to seek agreements within the system and because each decision made in response to the state of things as they are in that economic system is in response to a natural and undistorted expression of economic values. Each value so stated is in response to how things are and to how individuals assess them to be, and then choose to act. When impersonal and efficient, that assessment becomes easiest and most accurate and how individuals choose to act is then most realistic. Whether their actions are correct or not then are not the judgement of others or the system but of how their personal circumstances materialize in reality in relation to their actions. Then, the resulting economic system most resembles the human reality it is serving and, then, the resulting market exchange system is a social mechanism that is most consistent with the definition of man as a free and conscious human mind in the universe.

Thus, in the final analysis, we can see that the social mechanism of market exchange is the system of interhuman action that is most consistent with our human definition. It functions on a simple principle of agreement and exchange that is consistent with how we materialize our being in our reality. When brought to its simplest denominator, the values of our interhuman activity are made most evident in a market exchange environment that respects the Law of Agreement. Thus, to help make the principles of Habeas Mentem more our social reality, it is in this direction that we must turn our attention. From those observations will be the future directions that will need to be explored to bring the principle of universal order into our social reality, into a society of free and conscious individuals.

Chapter Twelve

In a Mechanism of Exchange

IN A MECHANISM OF EXCHANGE, as in a market place, the value of things depends upon what individuals are willing to pay for them. Things have a value either because we desire them implicitly or desire them because they are useful to us over time. The more we desire them, the more dear they are to us and the more we are willing to sacrifice in exchange for them. We will sacrifice and surrender more until acquiring them becomes too costly and our desire becomes overshadowed by what we need to sacrifice. Then we either curtail our desire or wait for the price to drop. Somewhere below that level is the natural price of things, at each moment of time. In the aggregate, it then becomes the prevailing market value.

It takes only two to form an agreement and, thus, only two to arrive at an agreed upon price of exchange. It takes a third to make it competitive and to introduce the potential for a higher bid or lower offer. Not all trades need to be arrived at competitively. It is possible to conceive of two individuals whose knowledge of value in each trade is such that they always exchange at an optimum price. Each exchange, theoretically, could occur at a price that is neither too high nor too low and, thus, not encouraging to others to enter in with a better price. If one has the knowledge of all the circumstances at each moment of time surrounding a potential exchange, then it may be possible to always exchange at a price that cannot be improved upon by a potential competitor. Realistically, this perfect knowledge is unlikely to be sustained over a period of time, but it serves to illustrate that at any period of time it is possible to exchange at an optimum price even in the absence of competition.

However, of greater consequence is that, since none of us can individually sustain this perfect knowledge over time, it is not likely that we could intellectually arrive at that optimum or, in effect, real market value. It is a real market value if it cannot be improved upon and, given that this improvement upon is subjective in nature and as evaluated by others in the aggregate, it is difficult to always arrive at it independently of others. Con-sequently, it is more likely that, in each trade between two in-dividuals, there is room for a competitive offer from another that would narrow the gap between the prevailing price being bid and that being asked. In this manner, competition can be seen to work as a mechanism that affects each exchange at the limit in between where the prevailing bid and ask narrows in-to the price of each trade. Over time, it thus arrives at the price that is the most optimum under those conditions that are the reality then.

Through time, the reality that defines the then value of ex-change also changes to suit the subjective and constantly shif-ting tastes and desires of demand and the existing and chang-ing ability to satisfy these demands. These changes, which result in constantly shifting market prices, are difficult to appraise without the entry of competitors eager and able to gain an ad-vantage from the opportunities created by them. Thus, the ad-vantage of a competitive exchange over one that is always de-cided upon by two individuals is that it allows other individuals to enter into the constantly shifting state of exchange and im-prove upon it to, in effect, contribute additional individual assessments of market conditions as they are then. The reward of this introduction of competitive assessments is that those assessments, which are most correct in relation to demand and how that demand can be most realistically and profitably satisfied, are those that will result in exchange. The price at which this exchange will take place is, consequently, that price that is most competitive and that cannot be improved upon, as assessed by all the participants involved and reduced from their individual decisions. Because it is the best possible price, or at least because it tends that way though it may not be perfect, it is also the most efficient price in relation to how things were then and to how the mind assessed them to be. An exchange is always a subjective human act judged by those individuals directly involved in it either by virtue of their demand or their ability to satisfy this demand; it is a price arrived at that is most relevant and correct then and there, to them. As in the case of our original two traders, because our knowledge is not always perfect, the competitive market environment tends to fill in those gaps that result from our failure to perceive all in our individual assessments. What we cannot see can be seen by another, as seen in another way or from a more advantageous perspective and acted upon to correct upon our unintentional omission. That other, either through being more clever or better positioned, then contributes a price of exchange that is more advantageous and, thus, more an expression of real market value. Thus, without the need for superhuman intelligence, the competitive market is able to improve upon our individual shortcomings and arrive at a comparably efficient price at all times. Its constantly changing price then reflects the constantly changing aggregate of human decisions as these are made in response to their individual realities and to how these realities seek each other in agreement. The result is an agreed upon and correct state of exchange.

It may be human to err, but in competition, collectively, we tend to err less. The advantage of a market system participated in by many individuals is that it has the advantage of many minds contributing their best decisions. These best decisions then result in the prevailing exchange. If, after the last exchange had taken place, the next bid and ask are uncompetitive, there will be a vacuum that, if possible, will be filled in by another ready and able contributor to narrow the gap between buyer and seller. If it is advantageous to do so, if it is both practical and profitable, then either there will be a higher bid or lower ask; if it is disadvantageous, then there will be none who would enter such exchange without loss. Whether a profit or loss is gained from an exchange is always determined by whether or not it is advantageous to close the gap between seller and buyer and thus to enter that transaction, at that time. The market only reflects the state of things as they are and the state of human assessments of these things at that moment of time. It does not punish or reward; what appears as punishment and reward to us is entirely the result of our action in relation to the state of things as they then are. Thus, if there is a particularly great risk in advancing a more competitive price, then the risk must be assessed in terms of the commensurate profitability or reward that can be expected from assuming it. Risk is a cost of exchange and competititon in the market system is not a free gift; it comes at a cost. That cost must be assumed in each transaction and is the level of risk that each must be willing and able to assume. For assuming this risk, the reward is a profitable return from exchange. In a competitive market, success gravitates there.

Thus, a market system naturally gravitates towards competitiveness. Unless a trader is such that he or she can always outbid the competition, the advantages left open in each exchange will always tend to be filled by another. Though it is possible, if not probable, for two in exchange to always arrive at the best possible price on their own, it is more likely that this best price would occur naturally in a competitive environment. Then, the market reflects best the state of things as they are then, there. How we trade is a reflection of how we assess each situation and how we choose to act in it. It is also an assessment and action that is human and localized in terms of the individuals affected by it. In a competitive market, we approach as closely as possible that state of things that best reflects us not only in relation to how we are but also in relation to how we are in relation to everything else in our society.

It may or may not be true that for the market to be efficient, it must be perfectly competitive. As shown above, it is conceivable that there exists an efficient market where only two exchange in agreement. Uncoerced, they can always arrive at a price that is relevant to them at that time; without competition, that price would be efficient to them. If the market system is noncoercive, nonrestrictive or approachably so, then all who wish to and are able to may enter it freely and compete; if only two choose to attend, then it must be judged that there is no incentive, under the existing circumstances of exchange, for others to attend. Competition does not have to be enforced; it forms naturally. What has to be enforced is the freedom from coercion that would seek to prevent willing and able participants from contributing to its exchange. When coerced, competition is restricted and the system works poorly. Perfect competition, then, need only exist as a potential insured by a freedom from coercion; as a prerequisite of market efficiency, it need only be perfect when the market strays from the best possible price. Then competitors must have the freedom to enter and fill the gap left by the prevailing exchange.

An efficient market system does not have to be an arena of many participants in perfect competition. That is a myth. On the contrary, it takes only two to agree on a price; a market is in fact most competitive where there are the least number of participants. Then, the trading environment is not crowded by competitors because there is no benefit from their being there; the prevailing price of exchange is optimum and cannot be improved upon: there is no advantage to be gained from it. A discrepancy in price creates its own competition; it does not have to be artificially maintained as a guard against market inefficiency; it needs only the freedom to be allowed to work as a system of mutual agreements. In fact, ironically, it is economically inefficient to force participants into competition where such competition is not justified because the price is already optimum and is not attracting competitors. To force such competition is to waste assets and human labor that could have been used elsewhere more profitably and productively. It is the mark of an efficient market when the prevailing price of exchange does not stimulate competition; if it cannot be improved upon, then to force competitors improves on nothing. Then, to force such competition is to coerce the market to act in a way that is not optimum, to make it less efficient, and to commit an economic error. Such is the power of myth, that it allows us to coerce where coercion is unnecessary and to force individuals against those agreements that are natural to them and that do not coerce others. When free from myth, conscious human agreements are free to seek their greatest profitability and the system of exchange becomes most competitive.

For markets to be efficient, they do not have to be composed of many participants, but they do have to be free from coercion. A market in which exchange is restricted, because entry is prohibited or because the costs of exchange are too great, is a market in which will not be reflected the greatest price efficiency. When free from this coercion, whether or not the price then reflected is optimum will be determined by whether or not the conditions of exchange are then optimum. If there is undue risk, such as from theft or currency instability or from confiscatory measures, then the price will also reflect the concern for this risk; the price mark up will be higher as insurance compensating for this risk. Then, if the price so arrived at appears to be less than optimum, it is only a reflection of conditions as they then are; the market cannot be improved upon if the conditions of exchange are negative. Exchange by agreement, when free from coercion, only reflects the state of things as they are between individuals. It is the property of free markets that, when allowed to work efficiently, they always reflect things as they are; if these conditions are constructive and unrestrictive, then they reflect efficiently our human effort and productivity; if they are negative and coercive, are plagued by undue risk and by disregard for the rights of the individual, then they reflect human inefficiency as forced from coerced labor. If we are not pleased with our results, the blame does not rest with the exchange mechanism; a free market reflects only human agreements. The correction of those conditions lies in correcting what the market is reflecting and not in correcting the market itself. If, however, it is the market that is being hindered from its free function as a reflection of agreements, then it fails as an efficient tool of interhuman exchange and as a reflection of things as they are; individuals must be free to form agreements. Coerced, it expresses reality only darkly and the myth that forces it to work poorly then becomes the new reflection of our social reality. That myth is then the attempt to change the reality of our human condition by forcing that which describes it for us; it is a form of social camouflage which masks what the agregate of our human agreements is telling us. Then, through our social error, the market ceases to be an efficient social tool. A market not free cannot be efficient.

In addition to the cost of risk, there is also a cost of entry into each market. If individuals, or their corporate extensions, wish to participate in a particular market, they must be prepared to pay that cost of entry. Simply, that cost might be a function of having to get oneself and one's product to market; this cost may not be great but it must be retrieved from the value received in exchange. Also, if the price at which exchange is taking place is already efficient, then it may or may not be advantageous to seek to trade in the prevailing market environment; the market efficiency may precipitate for us loss, a cost. Finally, the risks to be faced in future production or price or market acceptance may be too great to make entry into such enterprise advantageous at this time, thus unprofitable. If an enterprise and exchange is judged to be unprofitable, since it would cause loss, it should be avoided.

After such adjustments, if it is still deemed to be profitable to approach a market exchange, there then may still be the cost of operation associated with tooling up for the decided upon production. After the optimum price had been achieved, there can still be a kind of margin of profitablity in which prices may rise to a certain point above their theoretical or practical optimum without inviting competition. Competition would be held back because of the cost of tooling up production in order to participate in that market. Then, in that grey area above the most efficient price, is room for an enterprise to either enter the market and compete or to avoid the potential loss due to the cost of entry from tooling up. It is a situation which allows a price to be less than perfectly efficient, but it also is an area into which entry requires a keen assessment and a difficult decision. Because of that difficulty, the already existing participants of exchange then are able to enjoy a certain price advantage which affords them a somewhat greater profitability. Is that greater price justified?

Provided that the above competitive edge had not been achieved as a result of coercion; that is, provided that there are no restrictions on free entry, nor subsidies afforded by the either direct or indirect support of non-market agreements, such as taxes, nor that these goods so produced are non-exchange goods, such as roads or licensed channels of communications; then the competitive edge so achieved is a result of market action rather than either coercive action or action resulting from social agreements. In the case of roads or other services provided on the basis of social rather than market agreements, then the principle of market competitiveness and price efficiency do not apply; prices charged are agreed upon by the social agreement that provides those services. In a competitive market system free from coercion, the price that results is a product of market decisions in response to economic conditions as they are. Thus, under those conditions, what materializes in the market environment is a reflection of those decisions that had most successfully assessed the market reality. If it is decided that it is beneficial to tool up and invest the necessary assets and human effort into a particular enterprise, then the yield from the sale of those goods or services in the prevailing price of exchange must justify the effort. If the price is so close to its optimum level as to discourage this investment, though it may not be excactly at that level, then it is a competitive edge that is gained from the state of conditions as they really are. It is then too expensive to enter into this competition carelessly and too wasteful for a company or individual to expend labor and assets if the results will only yield a loss. Consequently, the decision as to whether this price advantage is justified rests on whether or not that competitive advantage had been achieved through coercion or through free market agreements. If they had been arrived at free from coercion, then they best reflect the reality of those costs associated with that particular enterprise and exchange, and to force competition under those circumstances in order to close the gap from a less than optimum price would only result in wastefulness. As we will later see, waste of assets and human effort is a serious detriment to society as a whole.

Thus, where human agreements are free to operate, there are natural safeguards against overambition and overzealousness, or carelessness. They are checked by what is feasible and profitable and what is foolhardy and wasteful. Free from coercion, the prices that materialize in each exchange are those prices that best reflect the situation as it is assessed by those individuals affected by it and as that market reality is relevant to them. When allowed to operate freely, it follows of necessity that what will materialize will tend towards that which is most efficient and most profitable and tend away from that which is most costly and unprofitable. To force it to be otherwise is to invite a state of things that are other than as they are and to force loss and human waste. It is an irony of market exchange that efficiency and competitiveness cannot be forced; they are naturally invited where they are justified; to force them to be where they are not justified only invites that which works most directly against them. In a society of free individuals, agreements formed in exchange are efficient only if they are voluntary and in their personal, individual best interest.

Chapter Thirteen

How Do We Measure Value?

HOW DO WE MEASURE VALUE? In each exchange, by what mechanics do we appraise the value of a trade in terms of something that is not the final object of exchange but an independent third factor, as an intermediary of exchange? What is an intermediary, or medium, of exchange? What is money?

When exchange reaches a level of complexity or sophistication that takes it beyond mere barter, it needs to employ a third factor that works as an intermediary, or medium, of exchange. This medium of exchange, money, requires only that it be more tradeable, more easily transferable in exchange, than would be the case of other goods or things traded. It needs to be more universally acceptable, to be more easily divisible, recognizeable as being itself, transportable, and non-perishable over time. Of the many things that could have been used as money, what has been advanced as money through time has been usually a metal commodity. Through a natural selectivity, the commodity that seems to best fit the above requirements is a precious, non corrosive metal, namely gold. Because of gold's properties rendering it easily recognizeable, divisible, consistent in quality, non-perishable, and, at a cost, transportable, it became universally acceptable as an exchange commodity useful in its function as an intermediary of trade, as a money.

With a commodity money, it became possible to transact a trade without having to acquire the desired end product directly, since it could be gained indirectly. In exchange for a good or service, depending on agreed upon values, was accepted an agreed upon quantity of gold. Later, in exchange for this gold could be received the good or service ultimately desired. Thus, money can be seen as a store of exchange value over time. Regardless of whether it is a commodity, such as gold, or agreed upon socially to be in some other form, it must be readily exchangeable and store value over time. Then, it becomes a convenient tool with which one can store the value of one's exchange until such time that it could be used again and transferred for some subsequently advntageous trade. In this way, all monies and currency have this in common, their basic principle: they have the ability to be universally acceptable and store value over time. If, for some reason, the money used fails to store value over time, it becomes exchanged for another money or commodity that can do so better; then it so becomes displaced. If this cannot be accomplished, the exchange economy returns to a barter system.

Now, imagine the following: that for each unit of money, regardless of its form, there exists in the economy a good or service which is valued by it. For example, if the money had been received in exchange for labor, then the products so produced by this labor can be, indirectly, said to be represented by that money. Or, if the money so received had been gotten through trade, then the products so surrendered for the currency received can be said, in effect, to be represented by that amount of money. This is only a mental exercise, but it is a useful one. Thus, it can be said that each unit of money, indirectly, is a representation of an equivalent unit of good or service available somewhere in the economy. It does not lay claim to it, it is not a direct certificate of ownership, but it can be viewed as an indirect claim, valued in terms of prevailing and agreed upon prices of exchange, on those existing goods and services. Because the claim is recognized only when there is an agreement of exchange, then money can be said to be a fluid certificate of representation of value in the economy. For each unit of good or service, there is a unit of money that corresponds to it, when it is agreed upon in exchange.

Now, imagine that the origin of this money, of a metal commodity in particular, started with the miner who had retrieved it from the ground. Its value to him, in still unrefined state, was less than to the smelter who had refined it; thus, the miner sold it to the smelter at a lower price than the smelter would resell it to a jeweler. But, in addition to the value added through smelting, the metal went from where it was valued less, at the mine, to where it was valued more, at the jeweler's. The jeweler could then resell it at a still greater price, if there is a market for his jewelry. In the case of gold, that portion of the metal mined that will not have usefulness either industrially or for decorative jewelry will find itself in a surplus valued most as a money. There, it will either trade within the value of its monetary function or be exchanged at a still higher price if it is desired for its industrial or decorative uses. As a jewelry or other manufactured form, it will remain there unless it should happen that the value of gold become such as to attract either more gold from the mines or to recycle gold from its other uses. Thus, unless the higher price of gold as money should warrant it, it would remain out of circulation and be formed either into a work of art, as in jewelry, or used for its industrial properties. Somewhere below that level, it would serve the economy in its function as a commodity money. Thus, while still in its primitive form as a commodity, we can already see money conforming to the principle where it can be seen to represent in addition to its intrinsic value, a unit of good or service for which it is exchanged; it can also be seen to flow from where it is needed less to where it is needed and desired more.

When money is advanced still further in a more complex market economy, it can be stripped almost entirely of its commodity value and become a unit of value representing its equivalent good or service available, through exchange, in the economy. At the limit, money can become merely an entry in a bank ledger, as authorized by the social agreement and as its exchangeability is insured by that agreement. Then, in addition to being represented as a coin or money certificate, it can also be represented as a claim against assets deposited at the bank, as a draft, and transferred from owner to owner merely through an account entry. Money need revert to its commodity origin only if this more sophisticated money fails in its function, as would happen if the social agreement that defines its quantity and insures its exchangeability is broken. Then, the market exchange once again reverts to a commodity medium of exchange, a commodity store of value over time. Thus, money can be any agreed upon unit of exchange.

Modern money need not be a commodity, unless the social agreement that insures its value is broken and it becomes devaluated; then it can be a commodity to insure its value over time. As a more advanced money, however, it can become purely what it is: a store of value over time; then, as a representative of goods and services in the economy, each unit of money can be understood to represent an economic good stored somewhere, or in the process of being created, in the economy. If these goods, through economic failure, should become scarce and less abundant, then their cost in terms of this money will rise; if the economy is healthy and productive and succeeds in creating more such goods, then such money can succeed in buying more per unit. As a successful money, it merely becomes a legal claim to goods and services, if and when they are exchanged, for an agreed upon price; then money is but an agreed upon representation of value.

Now, if money is a representation of economic goods, then it must stand to reason that how we spend this money is an indication of how we wish to direct or allocate those economic resources. If, for example, we spend a certain quantity of money on particular goods for the purpose of satisfying some human need, some need of consumption, then we are, in effect, directing which goods will be chosen for this consumption by what and how we buy; but we are also isolating those goods from the economy for consumption. The money that gets exchanged for those goods, through our choices, chooses which goods are to be removed from the economy and subsequently consumed. Thus, that money has acted, through exchange, to represent goods that are to be so consumed. It is hoped, under the circumstances, that these goods so chosen are not products of a static economy but rather are also in the process of being replenished through human productivity; otherwise, they being the last of their kind, their price would be forced up astronomically. Now, if on the other hand, the money in our possession is not spent on consumption but rather saved in a bank or invested in some other institution or asset, such as stocks or bonds or real estate, then it can be understood to represent an act where an equivalent amount of economic goods had been isolated from consumption and conserved for future use. This can be interpreted as an investment either through a financial intermediary, such as a bank or a fund, or directly by purchasing any asset or service that will continue to have value over time, any representation of capital. Then, what that money so invested in, or saved from consumption, buys also saves an equivalent amount of economic good and conserves it for the future.

Thus, of that money that is in our possession, either as as consequence of it having been received as wages in exchange for our labor or as the proceeds of an exchange, if a portion is not spent on consumption but rather is saved, then at that time an economic good somewhere or in some sector of the economy is simultaneously isolated from present use and saved for the future. This could be a piece of machinery or a tool or an amount of raw material, or unfinished product, or an amount of energy not consumed or an equivalent production of human labor, all reserved from being presently exhausted but reserved for a future use: they all represent a form of economic capital that had been conserved as represented by that money saved. It must be understood, however, that this savings cannot occur if the economy functions at such a primitive level that all must be consumed daily; then there can be nothing saved for the future and there can be no capital formation. Without economic capital to be saved, we would be forced daily, without tools, to gather or in some other way procure our food in order to survive until the next day. In a more sophisticated economy, investment is a natural act.

Thus, it is natural for us to isolate a portion of our income and apply it towards savings and investment. In addition to the benefit of our having a reserve for future use or as protection against the risk of future emergency, we also aid to isolate within our economy an equivalent amount of capital that will somehow be saved for a future use or for future production. How that money will be invested then rests with those individuals who have either directly or indirectly been entrusted with that decision. If that money belongs to an entrepreneur whose success allows him to reinvest his income or profits, then the decisions of what equipment or materials or labor will be purchased with that money rests with him. If it is deposited at a bank and reinvested there, either through loans or through purchases of investment assets, such as corporate stock or bonds, then the decision of what will finally be purchased with that money rests not with the original depositor but with the final user; he or she could be an industrialist or buyer or farmer or miner, manager, or whatever. The merits of that money's final investment will have been decided upon through many levels of successful agreements and exchange, in effect, collectively by all the individuals who had been involved with it. In any case, for each unit of money saved from income or from profits will be isolated in the economy an equivalent amount of goods or economic resources that will become, until consumed, capital.

In a dynamic economy, as opposed to a static economy, the state of the real market is characterized not by the state of being but by the state of becoming. Same as money tends to go from where it is needed and desired less to where it is needed and desired more, it also tends to go towards where it will be needed and desired more in the future. The future implies risk, and it is a natural consequence of a real economy that this risk must show up in the market as a cost. What things cost has built into it the risk of what things will cost in the future. What a farmer or miner will receive for his goods today, before the goods are sold, may be different from what they may receive in the future, when they are sold. There is a risk that today's costs of production may not be covered by tomorrow's price in the market. To correct for that risk, it would be natural to raise the price received today. This higher price, however, would represent less value, a lower purchasing power of money. The competitiveness of the market may narrow price within a certain band, but no closer; it would have to leave a gap for risk. Thus, risk can be seen as an economic cost that would go up with the greater the associated risk in exchange. If this risk is so great as to virtually prohibit transaction, then that economic function stops.

Investment of any kind implies risk. What is isolated in value today may not retain its value tomorrow. Then, it may no longer be useful or productive or profitable under changed circumstances. The decisions associated with evaluating risk require a unique intellect, almost a talent. Not all savers are suited to becoming investors; it is sometimes advantageous to pass on the responsibility of investment risk to those who are both willing and able to assume that risk, while one's own funds remain in a relatively safe vehicle, such as a savings account. A farmer or miner may not wish to stake his income on the vagaries of future demands for their products; it may be more prudent for them to avoid risk and hedge as much as possible on a futures exchange. Then, the risk of exchange can be passed on to those speculators willing to assume that risk with hopes of future gains the hedger of necessity forfeits. Then, if the speculator either makes money or loses it is of no concern to either the farmer or the miner or the processor; they are free to perform their functions relatively unburdened by risk, at a lower cost. What happens to the speculators, whose assets are committed to this risk, ceases to affect the economy as a whole, since the otherwise necessary higher price had been assumed by their speculation, and the successes or failures of their ventures gravitate entirely around them. Then, in exchange for their labors of assuming risk and making decisions in relation to these risks, they earn income through the transfer of these assets from those whose decisions are valued less to those whose decisions are valued more, from those who lose to those who gain.

When speculators enter a market exchange, whether it be a capital market or commodity futures market or money market, they enter it with the understanding and implicit agreement that they may face either gain or loss. What will be gained may be at another's expense; what another gains may be at one's expense. These are understood implicitly and agreed upon by the fact that they accept participation in such risk of exchange. What the speculator hopes to accomplish, in addition to assessing risk correctly and gaining from it, is also to take advantage of any price aberration and profit from that; they seek to profit from any discrepancy in another's judgement and assessments. This is a skillful function and one which helps maintain best value in exchange. But, most of all, they hope to amass, from their correct judgements and good fortune, a greater fortune of wealth than that with which they had started. They wish to make money, and that is their original purpose for being there. The rest only follows.

If the future were known and if markets were free of price discrepancies, then speculators would not need to exist; because they are not, if there were no speculators, their function would have to be invented. Without the professional speculator, it would be necessary for all individuals to share in, or pay for, this cost of risk. This universal assumption of risk, in itself, may be harmless, if it were arrived at without the need to force those individuals able and willing to assume risk from doing so, and if those who did not wish to did not have to. Because there are those who would rather not have to assume risk and actually shrink from the discomfort of that responsibility, it is best that this ability of investment and future decision be left to those who seem to be particularly qualified for it. Speculators thrive on it.

In the speculator's sector of the market, there is a natural process of selection of those speculators who are most successful; they are the ones who make money. The money they make comes, of necessity, from those who had, in effect, agreed to lose it, the other speculators. The general level of wealth committed to the market's speculative activities may or may not rise, but it is generally transferred to those who assess value best from those who succeed at it least. However, there is a rotation; not all individuals who are successful speculators remain forever in that enviable position, unless they use their wealth to coerce the market to suit their purpose; then it must be proven that they are so guilty of this coercion. However, that position of success normally goes to those individuals best at assessing and taking successful action in the face of risk; then it is those same individuals whose decisions will most affect market costs and influence the future course of the market direction. If not at the limit, their decisions will tend to become, through the process of selection of correct decisions, the future course of both the market and ultimately the economy as a whole. If it is they who lower economic cost through successful evaluation of risk, it is also they who, through their successful decision of future events, will indirectly influence the shape those events will take; they will direct their money, and the corresponding economic goods represented by this money,in that future direction. Speculators cannot influence the future; how individuals act in the future rests entirely with that population of individuals. But they can help the market and economy move in that general direction as they assess it to be. They are the spear point. The fact that they can do this well attests to their market success; they gain wealth. If they fail, they lose. If, because market risks had reached such enormity, they fail consistently, then society suffers loss as a whole through lower value. But if uncoerced, free to assume risk, they tend to perform their function well. If done by agreement through the mechanics of exchange, if free from coercion, economic value is gained from the resulting lower costs and society benefits as a whole.

Money, as do assets, flows from where it is valued less to where it is valued more. How this value is assessed, both in the present and into the future, depends on a mechanism of exchange where individuals are free to exchange both for the present and into the future. Then, depending on how they exchange will depend how economic resources will be allocated within that society. The more productive the society, and the lower the costs associated with the risk of exchange, the greater the value that will be defined by that society's money. Then, what things cost in that society becomes indicative of the values the individuals of that society had placed on their economic goods and resources; they become a reflection within that society of things as they are agreed upon, as they are. When free from coercion and free to create, values increase and money goes from those things we value less to those we value more.

Chapter Fourteen

Wealth is a Conscious Act

WEALTH IS A CONSCIOUS ACT. It must be chosen. It is the creation of the human mind and it cannot come into being of itself. Wealth is what is relevant, is the product of creative choices and is how the universe materializes in reality the values of those conscious choices, there. When what we do, in terms of human value, is relevant to us and tends to increase in value, we create wealth.

We can all create wealth. When we do in such a way as to build or to decorate or in some other way to form a thing of value to us, we create wealth. But, in order to be wealth, it must be recognized as such. It is not wealth if it has no human meaning, if it is not relevant to us in our mind and our being. If what we do has no value to either ourselves or to anyone else, then it cannot be said that the universe is materializing itself with value there. Then, that action is being done for itself and, in terms of wealth, is meaningless. If what we do only serves to lower value, to destroy it, then it can be said that the universe is working to destroy wealth there. If we have the power to create value, wealth, we also have the power to destroy it. How we choose to either create or destroy determines whether we can or cannot have wealth.

We cannot be given wealth if we do not have the power to earn it. To earn wealth is a conscious and productive action. The first condition of gaining wealth is that we are able to recognize it. If we do not recognize it, then we are not able to use it wisely or preserve it for the future and we are in danger of mindlessly consuming it. Consumed, it ceases to be wealth. Also, wealth must be exchanged for either another value of wealth or an ability; it must be earned. If we do not have the power to earn wealth, then we do not have the power to maintain it, which will lead to its gradual deterioration and ultimate decay. If we can both recognize wealth and then have the power to earn it, that is, that we do not instead consume it, then that wealth, as it is being used or consumed, is also being replenished and the level of wealth can either remain constant or it can grow. Then, we have the presence of mind, the consciousness, with which to use wealth wisely and with which to recreate it as it is being used or lost.

Thus, individual wealth is a state of mind. It is how the mind is able to create in relation to its identity, its definition, within its reality. The more conscious the mind and the more it is correctly positioned within its reality, the more it is free from coercion and the more it is itself, the greater is its ability to choose and create wealth. How a mind is in the universe, according to its level of personal development within its identity, so will be reflected in its reality that level of development in its personal wealth. What will surround it in its personal environment, in its possessions, will reflect that developed state of mind. If it is simple, plain, only ascetically aesthetic, then it reflects that personality; if it is lavish, ornate, materially abundant, then it represents a personality reflected by these. Whether primitive or highly progressive, or whether plain or gaudy, it is the reflection of the mind as it chooses its reality. On a social level, how that mind chooses in its society, how it is free to choose and seek itself within its reality, in agreement and free from coercion, will be reflected how that society will materialize its level of wealth. Then, a free individual will materialize in his or her personal social reality that same value that is defined by his or her definition in reality. If the reality is graceful, creative, beautiful, then so will these be reflected in its social reality. If that individual is instead lowly, destructive, coarse, then so will that personal poverty become the social poverty. Social wealth is the state of the society's individuals.

Thus, wealth is a product of human creation. A society is wealthy because the individuals who live in it are creative, hard working, productive, and because they have an appreciation for that which they are doing. They have respect for one another's property, they do not seek to destroy all they touch, they do not take by force, are honest in their dealings, and they seek to better rather than to trample and break apart. How they do this is a reflection of their state of mind as a whole society; how they succeed, regardless of their level of technical achievements, is a reflection of their level of civilization. A society is wealthy because the individuals within it care for their own and each other's personal realities, their human value; it is also wealthy because they are free to do that which they can do best.

A social order will always create wealth if the individuals within it are free to do so. Whether this wealth will be in the form of simple wood or stone carvings or whether it will be a highly technical and organized form of social capital will depend upon that society's orientation and the level of its individuals' collective development. If it has the ability and the resources, in addition to the freedom, it will tend towards those things that are made possible by civilization; if these are absent, then it will be immersed in the consequences of its own manifested reality. If they are dynamic, then their civilization is an active and creative society producing wealth; if they are passive, then their civilization may be peaceful and idyllic. What they choose will materialize as their social reality. Whether or not we approve of it is then but a personal value judgement; free, they can become in their society as they please. The successes or failures of their society then can only be measured in terms of what are their social expectations. These successes or failures will be determined by their social mind and the freedom with which this mind is able to materialize is reality.

When a society of individuals rises in sophistication beyond the level of each individual's self sufficiency, it of necessity develops that mechanism that allows individuals to trade that which they personally create. Individuals are then able to exchange, they can bypass their immediate personal needs, with what they create from their personal labor in exchange for what it is they need or desire. Consequently, it is then possible to devote one's abilities towards producing something for which one has no personal need but instead, through the market, for which there is a great need from elsewhere in society. If one can produce this efficiently, so fashioned from care and skill as to make it most desirable, most competitive, then one is producing from one's abilities in a way that is most desirable and beneficial to society as a whole. It is possible that one individual work on a part, that without the work of another individual on another part with which can be formed a whole, would be in itself valueless. Through the market, however, these two parts join together into a useful whole, it becomes a useful and valuable item. Thus, separately, their ability to create wealth is limited, since neither piece without the other is of significant value. Together, when these are joined into one, their final product is one of much greater value and, consequently, of greater wealth. Such is the power of exchange through agreement. Markets join together what can become of greater value than what had existed of lesser value separately.

Because of free markets, individuals are free to concentrate their efforts and skills on those things which they are best able to produce, in which they are most productive, and from which the market society will give them the greatest income. They will tend to work towards where their labors will yield the greatest value to them. If the individuals are craftsmen, then they can produce what may approach the finished product and which will approach that product's final market value. Then, on what they labor will be on what will be received their income as that product is sold. If, however, the labor is on what has no market value, in itself valueless without being used with something other, such as a part of a machine, then that labor would not be considered productive without the combined efforts of others. The part so produced is contingent on the cooperation of the labor and productivity of others in their specialized task. This could apply either to an assembly line employee fitting a particular part on another, or it could be an executive entrusted with the responsibility for decisions within a certain area of operations; either individual needs the cooperation of the other to be rendered productive in his labor. Without the other, each person's efforts yields little value; together, they form a valuable, if marketable, product from which the proceeds of its sale will pay for their wages and reflect their individual level of productivity within the whole operation. Each contributed to that value according to his skills and labors, and each received from society as those skills and labors were needed and desired there. How those were valued were then the derivative of those judgements and agreements that arrived at their value then and there, in terms of the state of everything else. Labor applied towards where will be gained the greatest income will then tend to be the most productive.

In a sophisticated society with a developed market exchange system, economic wealth is a function of how we choose to create with, and then allocate the products of our human and physical economic resources. How we do this reflects on how we create our wealth and in how we subsequently value that wealth in the aggregate of individual exchanges, the market. How we apply our income towards investment determines how we allocate those resources towards the formation of capital, and how we apply our labor with the advantage of that capital determines how we succeed converting that capital into our social wealth. Then, how we succeed in this creation, how this wealth is relevant to the other members of our society, determines the value that society will place on this wealth, which will determine how we will be individually compensated for it. Individually, though what we had created through our labor and skill had little value to us, because it had greater value to society, we gained an income that had greater value to us. Then, the enterprise can be judged a success and the value received for our productivity was meaningful in terms of how it benefited society as a whole. As opposed to the mechanics of a primitive society, in a sophisticated society it is not necessary to spend one's efforts entirely on a product that will have value only in its finished form; rather, one may be more useful to society as a whole and more productive in terms of the income received and what that income will buy if the labor expended is more relevant when it is combined with the labor and productivity of others. Then, we are most productive and are able to create a level of wealth that would be impossible if the same efforts were expended in a more primitive society. But this must be chosen; it cannot come about by itself and it may not be the most desirable state of affairs in all societies equally. For a society to reach that level of economic sophistication, it must be conscious of its production of wealth and how it has achieved it; otherwise, it will quickly dissipate it.

When individuals are free, they choose. What and how they choose then becomes the level of activity that is their society. Not all societies need choose the same production of wealth; other than perhaps for a feeling of economic independence and self sufficiency, there is no virtue in producing the same products that are produced elsewhere. When the world markets permit it, when they are free to operate and are free from excessive costs of risk or coercion, as in war, then what nations can produce together may surpass what they can produce alone. What they gain in compensation is then judged by how the world's markets receive their products, and how the world receives them is contingent upon how their products, at their prices, compare with such products produced elsewhere. Then, individuals are free to choose how to best produce and how to best exchange; together, jointly rather than singly, nations can then create a level of world wealth that would not be possible without their joint cooperation.

If, for example, a society is rich in natural resources and is able to sell those resources to the world, then it can contribute to the world's wealth what would not have otherwise been available. The income from these resources could serve that society either with economic resources for reinvestment or goods for consumption. How it decides the use of its income becomes entirely a function of how its individuals appraise their future needs and how they need to satisfy their present demands. If, however, a country has no resources, then, unless there be some severe social handicap, such as social crime or tyranny, it has the ability to provide human labor. The more sophisticated this labor, the more helped it is by a progressive level of manufacturing capital, the greater will be its productivity and the greater will be the value it can produce in the world's economy. Then, with the income from this labor, it can choose to either enrich its social level of wealth through a still greater development of productive capital or it can spend it on its immediate needs and pleasures; that choice, again, is entirely left to that society's social mind. If it chooses to create wealth rather than to consume it, then it will.

There is still one other extreme. It is where a society is neither rich in natural resources that could be sold to the world markets, nor endowed with productive labor competitive with that of other societies. However, if it is rich in scenery and natural attractions attractive for tourism, it then has a form of capital. This may be less than expected, since the tourist services may be provided by enterprises that are from abroad of that society. If so, then the proceeds of such tourism may not be employed back into the host country and be repatriated back to the society abroad. Yet, it would still induce local employment and generate income in the local economy. If this enterprise is attractive, then it will attract more capital, both from within the host country as well as from guest entrepreneurs. Though what may in fact be spent by the tourists visiting these natural resources of beauty may be but a small fraction of the income spent by them locally, if much is repatriated abroad. However, this would be one way of measuring the value of a tourist market by how much income it can generate in relation to goods and services that then can be purchased by the members of that society. In terms of world income, if a society has no more to offer than its natural attractiveness as a desirable place to visit, then so be it. It is its contribution to the world's wealth. If that should in time induce investments in other areas, such as manufacture, then its value as a magnet of tourism is then but a temporary value. Nothing need remain the same when it is free to change. But this is but a mental exercise.

What all societies share in common, regardles of their level of economic sophistication, whether a highly productive industrial and technological society or whether a tropical island paradise, is that what they each have to offer can have greater value if in cooperation with world markets than if produced only entirely for domestic consumption. As in the case of individual productivity nations can produce more in relation to the world's wealth if they trade and concentrate their efforts on developing that in which they excel and from which they can gain the greatest income. Then, whether or not they succeed becomes a function of how free are the individuals within that society to take advantage of that expanded market. If the world markets are free to materialize those exchanges, then the world's economies will reflect a greater level of wealth. If they are coercive, restrictive of trade and punitive towards individual agreements; if the internal national reality is also coercive towards its citizens and disruptive of their creative efforts; and if the reality is also such as to force coercion, as in war, the disruption of world's markets will be harmful to all those who had learned to rely on them. Then, individuals within productive and free nations will seek to trade only within like environments and avoid those that are unfree; the loss felt is most severe in those countries whose reality is one of coercion rather than of cooperation by agreement.

Wealth is an act of creation; it cannot come from social disorder. Whether this disorder is caused by a fierceness of a society's coercion, or by laws that encourage trespass or by attitudes that punish creative and productive labor, then such a society's ability to create wealth is severely hindered. Its population remains poor, perhaps with wealth concentrated only in the hands of a small privileged class which had gained it by confiscation or other coercion. But that is not the natural state of things; it is the least natural state in a society based on the principle of exchange by agreement. Wealth cannot be created from corcion or theft; it must be created from human ability and uncoerced human labor. In a world where value is exchanged by agreement, the wealthy person is he who is best able to create wealth and who is most instrumental in contributing this ability to society as a whole; it is not he who is most a thief, since that is the method to wealth in a society of coercion. Unless wealth was gained by force, and it can be proven so, then that wealth was gained by means of agreement and as that agreement had materialized around that person's creative and productive ability. Then, how that ability materialized and was accepted by others is how it was valued in exchange. The more creative and productive that ability, the greater that person's level of personal wealth, and the greater the level of wealth in society as a whole. To then take from that person his wealth by force, against his agreement, if he is innocent of trespass, is of no social benefit and is harmful to the individual from whom it is being taken; it is of benefit to no one since, then, the thieves unconscious of its value will quickly dissipate it. A society is not wealthy because it is most coercive; then, it only consumes capital. For a society to be most capable of wealth, it must be least coercive.

Thus, the society that will be most wealthy is that which can best shelter its individuals from coercion and which has the most free access to the world's markets. The agreements that describe its social reality need not be identical to the social agreements of other societies; they should be a natural reflection of the collective identity of its people. What must be identical, or at least consistent, with other societies if they are to be successful is that its citizens are free to form agreements and, if they choose to do so, are free to choose wealth. Then, wealth as a social product will follow of its own virtue. A society's success in wealth will not be a function of its ability to coerce others, either within or without its social borders, but rather its ability to be free from coercion, both internally and against foreign invaders. Then, internally, its individuals will freely form those agreements and values that are most relevant to them; externally, they will be most able to create and contribute to the world's level of wealth. Free of inner and outer guilt, capable of its defense and conscious of its strengths, such a nation becomes most conscious of its identity and most solid in its presence in the world. Then, it becomes the symbol of a free nation, powerful because it values individual freedom. Its citizens are free to pursue their personal identity and free to preserve that which is most valuable to them. Conscious of this freedom, they gain the courage and the ability to materialize the value of that freedom, and they become a nation of great economic strength. Free, they then can choose to rise to the level of the world's conscious nations. Wealth flows there.

Chapter Fifteen

In a Society Conscious of Itself

IN A SOCIETY CONSCIOUS of itself, it is the individual conscious of his or her identity who rises to the surface and gives that society its greater definition. When a people are free to interact by agreement, and are sheltered by law from being forced against disagreement, there will rise from amongst them those individuals more conscious and more successful in reality than the general population around them. They will be that society's leaders in all aspects of life and will be the ones who will lend it its special character. They will be the product of a natural selection of real excellence, not as determined by man-made rules of a man defined order, but as defined by that order that defines all things. How they are allowed to rise to their natural level of personal excellence and social leadership, to a high mindedness and spiritual consciousness, will define their greater society's level of greatness. In a society of Habeas Mentem, those individuals most conscious of their identity and most conscious of their social order as a definition of their consciousness, of a free society, will be the natural guardians of that society. They will be the leaders of their society's future direction.

We have defined a universal order based on an idea that can think itself and shown how, at its interrelationship totality, it defines each human being in terms of his or her mind. As is each person in the mind, so is that person in his or her space and time definition in reality, and so is that reality, from infinity, a definition of their identity, as is their mind there. When we become conscious of this, we become conscious of our personal real identity and begin to better occupy our greater

112

physical identity in reality. To better occupy this personal reality, we become more conscious of when we are forced from our mind, more painfully aware of when we are forced against our agreement, and become more conscious of holding our place in reality with a firm hold on the self. We have examined the social order that is best suited to this requirement of the conscious mind and the mechanics of exchange and agreement that make it possible. We do not have to have a social order that is identical to that of another society; it can be entirely unique in terms of how we have defined by our social agreements; but it does have to be based on a concept of agreements that allows the self to become expressed in its reality as its identity. This can be achieved only when the self is both free from coercion and is not free to coerce others. Then, the reality that materializes around that self, as that self is in the mind and as it is in its space-time position in infinity, is in the image of that person's identity in the universe. The physical reality, as it is fashioned by that identity, then becomes the social reality as it reflects the social mind. A person conscious of this is then conscious of his or her social reality and becomes aware of the consequences of his or her social actions. That person is who consciously guards society against the potential errors and trespasses of the unconscious mind. So guarded, society then becomes itself, as it really is.

It is for this reason we had examined the mechanics of a free society. It was to insure that the social order would not work such as to force an individual from his or her agreement with their greater identity. That falling out from one's identity is the greatest obstacle to our human progress; we could not know our social error until we could see that we have a universal identity. It was necessary to first realize a concept that defines man in terms of his or her definition in reality as that reality works to define itself; but it was meaningless to pursue this definition of man if this human identity was forbidden by the social order. Unfree, man was forced to occupy his or her reality not as defined by the mind but as defined by the social order; free, man became able to occupy reality as defined by the order of all things. If the mind and the human identity of the soul is greater than as defined by the social reality, unfree, the mind is forced to be less than itself and the soul is forced to slowly atrophy and ultimately die. We are creations of a universe at least as great as all that defines the colorful genius of the creative mind of man, and we cannot be prisoners within the confines of a social order incapable of materializing those human creations. If our human soul is to occupy its rightful place in its real identity in the universe, it must be free to do so, otherwise it will die. To do so to be free, it must be conscious of how its social organization will allow it to become itself. Thus, it must understand the mechanics of its society. If it does not, it will be imprisoned in a world void of the universal order, where the coercive actions of society work to negate their presence there, and where the mind of man is not free to be itself. With the power of the universe negated, the creative force that is life in reality becomes darkened, and the world is forced into darkness. We are conscious human beings in a living universe; when we become aware of this, the universe becomes more conscious of us, and reality materializes more its creative force in our world how we choose.

A society does not have to be wealthy for it to be free. We had examined wealth because it is a natural product of the conscious mind and the conscious mind should understand it. When it ceases to be a mystery, it can cease to be a social obsession and the mind can be free to seek itself beyond its material gratification. Wealth should be a byproduct rather than the product of all our efforts. In a world free of criminality, free of laws that coerce the innocent individual, and a world free of myths that act to destroy wealth rather than to perpetuate it, what is gained by labor can be increased multifold through machines, through productive capital; and what is earned as income can be augmented by the dividends yielded by that increased capital. Through a sophisticated technology, through an efficient mechanism of world exchange, the wealth generated should be distributed as an income not only in wages for those who directly contribute their personal labor but also in other benefits and dividends to those whose efforts and assets have put this machinery in motion. Wealth should approach that level where only a little labor applied should, through our machines, yield us a great income in proportion to the time expended.

How society then wishes to redistribute this wealth is then a matter of its social agreements; it need only observe that it does not coerce individuals against their agreement when it does so. But, though this level of wealth is imminent in our future world order, it is not in itself either an ultimate goal, nor is it a prerequisite for the society of a conscious mind. Society can be free and creative even without a great wealth generating machine. Not all need choose wealth, though those who do should be allowed to achieve it. The right to possess wealth is a condition of freedom that must be observed, but it is not the condition that enables us to become more human. The condition that must be observed to us to achieve that ultimate goal of becoming a true image of our human universal identity is the condition of social freedom itself. We cannot be creative and human without it.

It is this social freedom that is the responsibility of our social guardians. It is the responsibility of those individuals most conscious to preserve that order that is least coercive and that encourages the greatest consciousness. It is their responsibility to rebel if they find their society imprisoned and to establish the social agreement that will define their social order. But, in how they do this, what they must strive for above all, as conscious minds, is that they be consistent in their efforts with the principle that is their definition as conscious minds. That is their trust. Then, such a society most free can materialize in itself correctly in proportion to how is the mind in its reality. What is possible, what can be materialized in reality, if it is not forced on reality by coercion, is then what is correct within the universal order. What is impossible, what fails to materialize, is then what is being rejected by the universal reality. In this manner, the conscious mind can judge what it is that it is doing with the universe and what it is doing against it. Simply, what is in error is what fails to materialize in our reality because it is contrary to what is right in the universal order. To resist it and to continue to apply effort in this direction of failure, provided this is done without the use of coercion on others, will continue to yield no benefit and produce only a loss. When free from coercion, this occurs naturally and it is how the mind learns to communicate with its universal order, its greater reality.

When the mind can consistently succeed to materially yield its desired objectives and can do this with less error, the benefits of the universal order materialize within its reality. Such is the power of social freedom. The social guardians must be conscious of this, because it is their responsibility to make this power part of their reality and it is their responsibility to make the world free even for those individuals whose level of mental development has not yet allowed them to know this. Given the privilege of being treated as human beings, and being allowed to experience the consequences of their current reality, they will learn it in time. As the level of consciousness rises throughout the world, the gulf that now exists between those individuals who are conscious and those people who are not free in their minds will ultimately diminish and disappear altogether. Then, the world's societies will become more free naturally and the world will become more free to be naturally itself.

Thus, the conscious mind must learn to distinguish between what is myth and what is reality. Actions forced in the face of failure regardless of resistance are false. If reality consistently fails to materialize a given objective, then the methods applied towards that objective are in error. To perpetuate their application is then to perpetrate a social myth which only seals off society from its reality. It becomes negated by that myth from what is natural and twists into its place what is unnatural. We need not fear the universe; it will not betray us nor disappoint us. But, if we no longer fear it, we must respect its reality. The universe will not materialize for us what is against our human identity and what is contrary to its universal order. What is reality is what we can do, what forms through our creative touch, and what is possible through our agreement both with reality and each other. When we do not force reality through coercion, what we do is what is real and what is endorsed by the universal order; it is also what is real in terms of our human identity.

It is that simple. The society of Habeas Mentem can be reduced at its simplest to an equation: When we do only through agreement, when we obey the Law of Agreement, we are most free. When free, as we do, so we succeed in society only with what is possible and fail with what is impossible. What is possible is the action the universe materializes in our reality and what is impossible is what it rejects. Thus, how we do successfully when most free is how the universe materializes our social reality in terms of itself. It is how works the universal order within our Earthly reality in that free society. What develops from this society is then the entry through which we progress into our next level of human development and through which we enter into the universal community of conscious man. When that equation is made possible by society's guardians, society becomes as it really is itself.

Thus, it matters little what form is the society in which we inhabit. Some societies are very old, venerable in their identity by the countless generations who had lent it their spirit through the centuries. Others are new and bold and, if still largely untried in terms of history, are already successful in their creative contributions to our world's reality. It is not for the conscious mind to change these. These societies are real images reflecting the real mind of their individuals as a people. The colour and richness of each world is what lends the universal human reality such a wealth of spirit, and these should always be encouraged to be themselves. We can be different; we need not all be the same. There is no harm in this diversity of human identity if we can universally learn to respect ourselves and one another. Our reality is every bit as valid as the reality of another; if we are different, it is only a reflection of the richness of a universe that can be different for each one of us.

Nothing need change. The society of Habeas Mentem is not a radical, physical change of its social structure. Things go on much as they had done so before. The same institutions, social mechanisms, laws, the same social characteristics can prevail as they had before. Ours is not a mission to change the world, only to insure the inviolability of the free conscious mind; the world will change of its own when it is ready. All that need change is an awareness that man has a mind and that as he or she is in that mind, as it is defined within its personal real identity, is how that mind is in the universe and how the universe materializes its reality in response to that mind. What then changes is that man becomes aware of himself and of the definition that allows that awareness. What will change in society is where there had been cruelties, inconsiderateness, barbarism, and wanton destruction. These will increasingly become characteristics of a less conscious past not becoming of the new free individuals. Such awareness, once firmly established, takes firm hold of the social reality which cannot be relinquished by the conscious mind. A free individual knows why he or she is free, knows the value of this freedom, and does not weaken in the face of adversity when this freedom is challenged. A conscious mind knows reality and knows what is falsehood. It knows these because it is conscious of the fact that in the real world falsehood fails to materialize. That is the awareness of the new man. It is the awareness that makes society become truly human.

Thus, as conscious minds, we do not seek to conquer, nor to convert, nor to lie and corrupt. We do not seek to change the world. When it is ready, the world will change of itself. What we seek is that we be not trespassed against, that we have the right to be in our mind. To the conscious mind, to be trespassed against is intolerable; unfree, it cannot be itself. Itself, it succeeds in materializing a world whose success we have hitherto but had fleeting glimpses of during exceptionally creative periods of our history. We tend to judge past events of history by a chronology of conflicts and wars. In fact, we had progressed in spite of these and their drag on our human development. The human progress that had brought us forth from more primitive times into the present has been a history far richer than that of our human conflicts. Against sometimes heavy odds and great obstacles, the mind of man inched forward through the works of countless and obscure individuals. Through labor and intellect, they had precariously carried the world forward into the advances of the present. If there were sudden bursts of genius, these were but beacons to light their way; the genius is powerless without the endorsement of the minds and acts of many individuals. History was made not of great battles and great deeds but by those countless individuals who persisted in living through it all and who overcame their suffering each in his own way by his own ability. That is how is brought change into the world; it is only this form of change that is truly constructive and creative. It is not for us to convert the world; it is only to help make man as he truly is: then, the world will convert itself.

Nor is the conscious mind conquered; it does not submit. The social guardians are those who are forever vigilant of the safety of their world and conscious of the forces of their society. They tolerate the presence of all other societies, but they do not tolerate another's aggression. Aggression is an unconscious act, and it is the responsibility of the social guardians to guard against the trespasses of the still unconscious mind. A mind confident in its knowledge, and aware in its reality, is not afraid of aggression. It finds it loathsome, revolting, and acts directly to repel it. It repels it swiftly, efficiently, almost clinically with the least shedding of blood. War is not a tool of diplomacy, but war is a cruel failure of understanding, a failure of diplomacy. It must be entered into only if no other avenue is left open. Once entered upon, it must be fought with total determination and conviction and the full power of the free mind of man. It is not to be entered into casually or thoughtlessly; but, once engaged into it must be executed swiftly and efficiently. The unconscious mind is not heroic and will generally not attack where there is strength; however, if it should attack, then it must be repelled. The constant vigil of the conscious mind against those circumstances that could result in attack are a severe responsibility towards the well being and survival of its society, and steps should be taken far in anticipation of conflict to avoid war. But, if these fail, battle must ensue. Whether the war is then fought through vast military mobilization or through covert activities, it must never be fought halfheartedy but always fought to win. The aggressor will not seek to trespass into where there is material and spiritual strength, and these are the conscious mind's first defense. But if society is attacked, anguish and bitterness must be held back and, as a last resort against the unconscious mind, the aggressor must be destroyed.

Those are the hard responsibilities of the conscious mind. If the mind is to be free, it must be aware of them. It must never weaken in its self awareness, for that is its primary real defense. So protected within the borders of its society, it is then free to pursue its greater development as a sensitive reflection of the universal reality. As the mind seeks itself in its greater identity, as the world's reality begins to reflect more clearly the creative force of the presence of a newer, more conscious man, as man progresses in spirit, the ferocity that has characterized the state of the present world becomes subdued. The social wilderness becomes tamed by its greater human image within the universal order. The darkness that had filled the planet gives way to a new light, to a new "I", to a new "I am". What had at first been but a cognition of the self, and later had become a recognition of one's being, and to think, had now become the consciousness of one's being within reality in an infinite universe. The mind can think itself, become aware of itself in reality and think it, as the universe's reality is thinking of it. The spirit that had moved us long ago to say "I am" now moves us into a still higher awarenss: "I am conscious". "I am that I am". If our past world had been directed by only "I am", our new world can now become directed by a still greater: "I am the reality that is my being". "Free, I am in the image of the infinity that is my human being, in my mind". Then, the reality of our new consciousness, fully human, is the reality that has the power to propel us into the next level of our human development where we can say: "I am my human being". Then, all reality can become charged with the energy of our new consciousness, there, our new human identity, and what had been but a small physical world lost in a great universe can become a planet conscious of a universe become conscious of us. Then, more conscious of our new human being, each one of us, the universe becomes more conscious of us. In "I am my human being" we become new man, new woman, and our planet becomes a new identity within the natural order of the universe.

This is in our power to achieve. It is that simple, though it is infinitely complex. Reduced to its simplest, it is but a matter of individually reaching out for it, of choosing it. Can we achieve it now, or do we wait until we are more confident in our being? The answer is entirely ours. We need only to have the right to choose, the rest will follow. Whether we choose now or later will always be reflected by the universe both in ourselves and in our social order. Reality will work with us if we choose it, and it will work away from us if we reject it. The universe will always reflect how we affect it; if we do this consciously, free in our identity, it is how the universe will affect us. Then, we will be the universal materialization of a society conscious of the liberty of its individuals, a society conscious of itself.

"Who are we?" We are how we occupy our personal space in time in reality, our personal identity; and we are as a society, as a planet, how we occupy our space-time reality in the universe, our universal identity. If we fail to choose freedom and instead still negate our identity through coercion, then those forces that define the universal order within our social order will be equally negated and our planet will also be closed to the consciousness of other worlds. If we live in darkness, then none dare approach us, for life here would be unhealthy and dangerous. To reach out into the universe, we must consciously choose those values that reach out to us, that define the life of our being, those that render us truly human. If we reach out for freedom, for gladness, for goodness, for kindness, for compassion, for responsibility, for fairness, for honesty, for all those things we love and are beautiful to us, then how the universe reaches out to us wll be equally true and beautiful to us. We are on a threshold of an exciting and great new adventure as a human specie. We are on the verge of being accepted into a new dimension of reality where our vision can be raised to a new spiritual revelation, to where we can open our eyes to the vision of our soul's Creator. It is a big universe and we are a great people and what we have in us the power to follow can open before us great dimensions of our infinite Living Reality. If it is in our power to achieve this, it is not so strange. In our minds, we are the human beings. It is our identity. In the space-time relationships that are the interrelationships of Life, we have the mind. Imagine an idea that can think itself. We came from there.

Chapter Sixteen
The Mind is the Communicator

THE MIND IS THE COMMUNICATOR. What we think and do in our mind, when we are in the mind, is a reflection of what is our identity in the image that is the totality of our being, at infinity. Through interrelationship, all things are connected through space and time, through distance and spatial relations, and through events and circumstances; at infinity, that total interrelationship forms an image, an understanding closed in on itself, that defines every part of itself in infinitesimal detail. That image is ingrained into any part of itself that serves it as a focus, that is the focus from which the infinity is observed, its ultimate center. Infinity looks like itself there, then, its qualities minutely ingrained, condensed into the focus that describes that moment and its infinitesimal part, in terms of everything else, ad infinitum. When that part becomes aware of itself, when its consciousness defines it in terms of its position in the universe, that ingrained focus becomes a mind conscious of itself, in terms of its universe, and exists as an identity over time. Then, that mind will reflect in itself the image that is its identity at infinity in its reality. Its thoughts and actions will materialize its world in that greater image that is its universal identity. That is how the universe communicates itself into our reality.

That same power that became the mind, our consciousness, is the same force that struggles to become the focus of that mega-structure of man created by his mind, his society. For the consciousness of our universal order to penetrate the mechanics of our social order, we must be in the mind. We must occupy that infinite variety of possibilities that describes our conscious mind, as it is defined by our universe. Society is not a closed system dictating to the mind its responsibility within the social whole. It is more than that. Society is an open system that gives the mind the freedom within its greatest whole, to express itself in all the possibilities that are its rich and varied creativity, its real responsibility. To force the mind within a closed system is to force it from its time-space relationship and to stifle it, to cage it within infinity. A mind forced does not communicate infinity; it merely reflects the limits of confinement of its social cage. A mind enslaved is not a communicator but a tool of that dark force that restricts it in terms of its identity. That force, that coercion, is what closes society in on itself and limits expression of the richness of the mind as it is in its universal consciousness. Free from this forced confinement, free to be in the mind, the social order takes on the greater consciousness of the free mind. But it must be free. If enslaved, it becomes but the dark tool of those forces that work to close off the social system from its greater development within that infinite mega-structure that is our universe. Darkened, a closed society focuses and closes in only on itself. A closed society is unconscious in the universe.

The social order is a creation of the mind. It is formed in our image; its mega-structure is a product of our collective mind. But as all such mega-structures of the mind, it has the power to destroy its creator. The ultimate power in the universe is that power that is its own self destruction, chaos. All things can always return to chaos. Destruction is always easier than order; it never structures itself but rather triumphs in its lack of structure. It is the mind that lends order to where there is disorder, first by finding order, by defining it, and then by reordering it according to its definition. There is no freedom in disorder, for there is no agreement with it. It is impossible to come to agreement where total disagreement always prevails. It is the mind that finds the power to harness these and to form agreement. There is no social order where there is disagreement, only social chaos. Chaos is the ultimate primeval ruler of the universe and it had reigned supreme until came mind. But the mind subdued it; it spanned it infinitely into a universal order. Within this order, it created man as a mind with all the systems that can be compatible with the mind's Creator. But in the mind is also the power of chaos. Disorder is as possible as is order and man is always poised for his own self destruction. This self destruction starts with his social order which turns in on itself and then culminates in self mutilation. When the social order is a closed system, the universal mind is negated, and chaos is on the ascendant. Then, society ceases to be the creation of the mind and becomes rather the destruction that exists without mind. Without society, man simply returns to the primitive level of self survival, in a world of coercion where the successful survive through combat.

It is important to be in the mind. First, we must be conscious of being in the mind to become more conscious. Then, we must be more conscious to succeed over chaos. We must use chaos to explore all the possibilities that are the universe; the irrational is our ultimate freedom; but we cannot be subdued by it. In a closed society, where the mind is defined only by the confines of the social order, we are not free to explore our universe. What is not socially endorsed is forbidden and cannot be approached further. But the mind rebels within such confines. It seeks greater expression, greater creation; it risks chaos. But if it is overzealous and overindulges in the chaos it courts, it is in danger of destruction. Such is the balance of to be in the mind. To force, to coerce, to force against agreement is to perpetuate disagreement and to court disorder. There is no freedom in disagreement where chaos prevails. That is the limit of our free will. We can only do as we agree to do. To do otherwise is to force oneself outside the mind. Outside the mind, we lose its protection and guidance. Then, we are victims of chaos and society turns in on itself in destruction, and we turn in on ourselves. To be in the mind, our will must cooperate within the limits of agreement. That is how we become free. To be free is how our mind conquers chaos in the universe.

Our mind is the most powerful expression of universal order. We are not yet conscious of it all, but we are conscious enough to succeed in creating a social order. Thus, we are conscious enough to materialize in our social order the power of the super-mega-structure that is the universal mind. It is but a small reflection within a greater society and built but in its image. But that is an impossibility in our terms. We are not yet conscious of the mind of the universe; we are but vaguely aware of its existence. To bring this ultimate order into our own miniature social replica of that super-mega-structure needs that we assemble a system that is defined not by our consciousness, since this is still lacking in its entirety, but by the process of reality that defines that consciousness. We must simply occupy our identity in its space-time definition. Then, though of still limited consciousness, we can physically and mentally position ourselves in a way that aligns our minds with the identity that is our mind out there in the universal order. That is what is meant by an open ended society. The limitations on the mind are thus not man made, except where social laws restrict coercion and prevent one mind from trespassing on the reality of another, from forcing the other against his or her agreement; but, instead, are limitations imposed on the mind only by the conditions of reality and our existence in it. We then become open to all the possibilities reality makes available to us. That is how we learn. The mind becomes conscious enough to understand its reality and, by positioning itself to become one with its greater reality, it becomes increasingly conscious of it. Its awareness deepens; its understanding sharpens. Simultaneously, the social order becomes increasingly conscious as a reality, as a mega-structure of the mind and as a reflection of the struggle of the universe's order over chaos. By positioning ourselves in our identity, we move and are moved by the universe in a way that is more powerful than could be moved otherwise without the assistance of the universal order. Together, they restructure reality in their greater image.

To be in the mind is the principle that is the Social Contract of Habeas Mentem. To be in agreement with oneself and with one's fellow man is the greatest achievement of the universal order. It is to be true to it by being true to oneself and towards one another. In it in agreement, is the principle of the mind succeeding over the destructive chaos of the primordial. We are powerful when we agree together because that is when we are most free within our universal identity, when we are most conscious, and when the mega-structure of the universe works most closely with us. But agreement cannot be forced; it must be spontaneous and sincere and yielding. It must be sought voluntarily. That is the condition of our success of mind over chaos. Chaos destroys regardless of condition; agreement succeeds only when it is invited. It is the ambition of the Contract of Habeas Mentem to invite agreement into our social reality. Then, in agreement, free from forced disagreement, we invite the super-mega-structure of the universal order into that miniature replica that is our own society. In Habeas Mentem, when we will become conscious of it, we will ultimately individually have access to that same power that is in the mind of the universe. But this cannot be forced; it must be in the form of a contract, in effect, a voluntary agreement arrived at by our choice.

Thus, the first condition of human consciousness is that we choose consciousness, that we must seek it. To beget the benefits of consciousness, it must be voluntary and willed. Consciousness increases with the increase in its own volition. But it increases, of necessity, with an increase in its sensitivity. Consciousness is not coarse and does not destroy because it does not force against agreement, though it uses force. It does not corrupt because it seeks what is genuine in agreement and thus would not force an agreement on another. It is noble because it is sincere; insincere it would fail to be itself. It is sensitive because that is how mind finds order in disorder,by being open to all nuances. All these can occur only voluntarily. To force them otherwise only negates them and forces the mind further away from its goal. That is a condition of reality. It must be obeyed. A mind must choose consciousness or it will fail to be in itself and negate the universal order around it. When it obeys these conditions, that is how the mind becomes supreme. There is a natural order in the universe. Conscious, in our mind, free within our identity, we are in our will the natural order of the universe. We do have a free will, but it is valid only, and powerful only then when it is free to be in its image and then be in that image, because that is how we communicate with our universe.

We are the communicators. We create with the free mind of man that which is most human; we destroy that which is most unlike man. But we can do this only while man is in the mind. Outside the mind, we fail as communicators in our universe. The mind of man can be trapped within a social system that prohibits freedom of thought, of expression, of agreement. Unfree, our human message becomes unclear and ultimately corrupt to materialize stillborn as error that ultimately plays into the forever backsliding forces of chaos. If our thoughts, our philosophies, encourage us to lie, to steal, to twist reality into untruths, to revere in ugliness, to waste, then what will materialize around us will be equally twisted within the universal order. Then, to be in the mind becomes an increasing difficulty, if not impossibility, and only with great pain and determined effort is the physical collapse reversed and consciousness is again reestablished as the mark of the human being. Unconscious, we are less than human. Less than human, we fail as communicators. Only as earnest seekers of truth can we be communicators, and only then so do we project in us and around us what is truly human. In order for the universe to communicate into our Earthly reality, we must be fully human beings.

Thus, to become human, fully conscious in the image of our universe, is the greatest goal of our development. We are not in our kingdom yet, though we can approach our kingdom on certain conditions; ultimately it will be for us to rule. We have sought ourselves by casting ourselves off into space, but only to realize that the space around us is our personal reality, at every moment of time. We could use that knowledge to better find ourselves but there are conditions that must be oserved. We must reach out for it, though it is forever reaching out for us, but the mind cannot know it until it chooses to know it. There is work, it is effort, exertion, but freedom does not come with the ease of chaos; it must be sought with a distinct responsibility of agreement. There is a Social Contract to help guide us, but it is a guide effective only through our agreement. If we do not choose it, it will work on, but away from us; the conditions of the universe will always remain the same. There is a social order we have created, one that hugs us closely in our everyday environment and protects us from the hostile elements of a still primordial world, but it is but a ship we had built to guide us through a universe that is here still dangerously close to chaos. But we can now have the power to steer it. We still do not know it for certainty, but by being in the mind, we can harness the power to steer it through the power of super-mind to move us closer to being human. When human, each one of us will share in that mind and be a social contract unto himself. Fully human, that social agreement will equal the agreement that is our universal order. What is now our human conscience will develop more fully and become that force that will anchor our thoughts and acts in reality. We will no longer be ruled by fear but by the confidence that has learned to overcome fear. Then, the mind of man and the mind of the universe will be as one.

One can move the other; they can communicate. If the universe can move the mind, when consciously human, the mind will move the universe. A human action can then become magnified in its real terms, become enlarged in terms of its compatibility with the reality around it, and move all those interrelationships that define that moment. To be one with the universe is to have the universe become one with the mind. When consciously human, there is knowledge of what act is possible with the amplification of the interrelationship of that act in terms of everything else, and what the mind does then the universe does with it. It is like a power of magic, though it is a real force. It can happen now, if we choose, but still only on a social scale; then, we could have it happen on a personal scale; but first we must learn to be in the mind. Where the social order can be made to reflect the order of the universe, by being in the mind as a social contract, as a right, when fully in the mind, fully human, each person's reality can be in the image of its personal universal order as it is ordered in the Creativity of the universe. The mind that can now be moved by the universe, and still be but semi-conscious of it, can then become moved in a manner that is fully conscious and move the universe in return. We may already do this to some small extent, though we are not aware of it. Being alternately in and outside the mind, our results are sporadic and uncertain. At times we seem lucky, at times we are not. It can be that mundane, though in real terms such self gratification is of little significance. We are so much more than merely the self. We can move the universe to create in a way that it cannot Create itself.

We are the Creators. We are conquerors of chaos, we are users of it; but we are but mere communicators. We are the force that can change reality; but we can act only where we find agreement. We are builders of worlds; but we are but individuals. We are the human beings. Through us flows reality from its ultimate chaos to its ultimate order. We are the purifiers; we are processors of reality into minds. We are seekers and explorers, we are adventurers; we are creators, but we are but finders and definers of beauty. We are human, though we came from animal. We are dignified, though we have regained affection. We are fierce, though we have learned to be gentle. We are conscious. That is the beauty of man on Earth. We are creatures of our planet, but we have gained the mind of our universe. It is a precious gift which we must not surrender to chaos. We must not let it be destroyed in self mutilation. But we will prevail. Our universe will move us to move with our universe. But first we must become the conscious human being. We must have the freedom to be in, and to have, the mind. The rest will follow.

* * * * * * * *

HABEAS MENTEM

Man's mind, when it is most intent upon any work, through its passion, and effects, is joyned with the mind of the stars, and intelligences, and being so joyned is the cause that some wonderful virtue be infused into our works and thing.

Henry Cornelius Agrippa
Counselor to Charles the
Fifth, Emperor of Germany, c. 1530

BOOK II:

HABEAS MENTEM II, To Have the Soul: A Spiritual Reality.

Chapter Seventeen

How Do We Know — "Who Am I?"

H OW DO WE KNOW what is the soul? Is it described by the mysteries of ancient religion, the "ka" of ancient Egypt, the Holy Spirit of the Christians, or the re-incarnational soul of the Hindu? Or, perhaps, the soul is more like the "mind" of the ancient Greeks; or a daemon, every man's soul a guardian over his life; or is it more like the essence of "being" of Tao, or the meditational state of enlightenment of Buddhahood? Perhaps the soul is described by all of these, they all being facets of man's efforts at identifying himself in his great universe. The soul may be nearly material, like the near matter of an ectoplasm at a seance; or it may be pure being in its most abstract. Each of these are our efforts to contribute to our knowledge of the mystery we call Life, that which enables us to need, to yearn, to seek, to do, and to dream; to Love. Who are we? Who am I? What is this that we have that we would call a Soul?

The search for being that is called the soul has preoccupied man throughout his existence. Our ancient forebears, in a still primitive state, already were careful to bury their dead, leaving behind tokens of their existence to be used in their later life, consecrated with ritual and belief in some deep seated magic to ease passage into that other world. Man had groped through the ages for a glimpse of himself as a being greater than merely an animation that would span a few generations, a finite number of years, as his existence only to perish forever. It was not easy for the conscious mind to accept such total defeat, and the mind created itself a greater reality. Man would not die and slip away into a certainty of oblivion; he would somehow survive, gamble that he still existed and thus would not be

133

cheated of his existence by death. The vehicle chosen for this victory was the creation of the soul. Through the soul, we could pass on into the next dimension of our existence, whether this be a totally different being in a world beyond or as a return to this life in another body. It is a compelling idea, one that would survive the eons to the present. Even today, in our modern world of sophisticated agnosticism, even in atheism, the idea of continued existence survives. There is something durable to the idea of a human soul, something rooted deep in our consciousness that does not allow us the peace of ultimate certainty. We do not know about the soul, we are forever anxious over its existeence, but we would not die. We are persistently posed with the question: Who are we? What am I? Who am I?

We are now in what is to us the modern world. Tremendous strides have taken place in our socio-technical development within the past centuries. We can move about with relative ease and great speed, though we may not be free to go where we wish. We can communicate immense information at great distances within seconds, yet, in many parts of the world, we may not be free to use this information as we will, nor even listen to it. We can produce, trade, live in affluence that ages ago would have been ordained for only the most privileged, but many in the world are forced to live in abject poverty and without human dignity by laws that prevent them from being themselves. Much of the world is still characterized by corruption, oppression, and other criminal expressions where one individual will seek to gain from another at the other's expense. Brought to its acceptance on a world wide scale and the world is faced with world wide misery threatening its modern achievements.

We have fallen in a way that was inconceivable to our more naive past. We are insecure in our new achievements. The world had once advanced tremendously in its spiritual ideas, only to fall back upon the fact of its real, material poverty. The ancients felt secure in their ideas of their universe, they were often wrong in their conception of it, but they were secure in what it was they believed about themselves. This confidence in the order of things gave them a certain strength and dignity that carried them through their other shortcomings. Today, we have a reversal that has given strength to our material achievements but that has left our visions of ourselves sadly begging. We are insecure. We spend time psycho-analyzing ourselves but we do not know ourselves. We have overcome those ideas of our past that are odious to us: torture, exploitation of the weak, the young, women, races; yet we live in a world of chronic fear. One would have supposed that with the elimination of those ills that have in the past plagued society, with more of these social injustices removed, society would rid itself of its errant behavior. Yet, in spite of admirable achievements, we are plagued with terrible social and personal disruptions.

In our vast material achievements lies a terrible void. There is a need to progress in a way that is somehow spiritual. Our modern scientific sophistication has allowed us to scoff at the idea of a soul, or of a heaven in the clouds, or of a hell, or even of the existence of God. These are obvious vestiges of our more ignorant past when men were superstitious and naive. To revert back to those days in our thinking appalls us, understandably. Some have turned back and found comfort in their religion, or in the religion of other cultures, as in those of the Far East, or in rediscovering a spiritual simplicity as in a born-again movement, such as a new Christianity. But these are not the main force of our current development of progress. It is possible that a fringe ideology will eventually triumph and lend new direction to the world, but it is unlikely that man will revert in time willingly. Progress is a willed positive force of development that is chosen and exercised in a way that constructively moves reality. It needs a firm concentration of effort on our progressive course so as to not revert back to our ignorant, primitive past. The past was cruel; the present strives for enlightenment. The past had wars, tyranny, slavery, famines, disease. The present is still plagued by these, though to a large degree social acceptance of slavery has been eliminated and we have made great progress in combating communicable diseases. But we still live in fear, in the shadow of near instant destruction by our formidabe new weapons or of slow death by their byproducts. Social and political terrorism are rampant. We cannot even feel safe in what should be our monuments to our conquest over the wilderness, our cities. We live in chronic fear of sudden attack or rape or robbery by our fellow citizens. Within this existence of material wealth and spiritual poverty arises a certain hedonism that urges us to enjoy and spend and pursue our physical pleasures, to emulate the happy smiles on advertising posters or television, or to find fulfillment in their sensousness implied. We are flogged mercilessly because we are lost. In all our weath we are poor. It is no wonder that the perfectly rational mind, given these options, would select to believe in nothing and seek satisfaction in the material pleasures of the modern world that for a price are made so easily available.

Why do we strive so? We are forced to ask questions about ourselves, to seek our identity: Who are we? In Book I of Habeas Mentem, "To Have The Mind," we have seen how "being in the mind" is a definition of our human identity. But this was only a metaphysical definition. What does it really mean to be in the mind, in one's identity? We can know that through the mechanics of interrelationship we have a greater being in the universe that connects us with the rest of our existence and that characterizes the conditions of our physical being. We can know that to be in agreement with oneself and with one another defines that greater being in our environment and connects our personal mind with the mind of our infinity. We have also seen that when man is free to be in his mind, in his identity, he or she is more creative and is in how the universe communicates into his or her existence. But to be free is a conscious act, one that needs to be chosen through agreement to gain that precious foothold on becoming oneself in one's reality. Once chosen, the mind is then free to seek the meaning of its greater identity. We are at that threshold now. We can know what is meant in the Habeas Mentem metaphysically, but we do not know it in ourselves, intimately. If we have chosen freedom, then we may seek identity. That was the condition encountered in Book I: to reach this level one had to choose freedom, for the answer to "Who am I?" would be meaningless without it. Now, we can be free to resume our quest where Book I was forced to leave off in pursuit of social requirements for identity. We can now transcend social inquiry and seek personal consequences of our new identity. Thus, we are now free to entertain the existence of our soul. Socially free, in our mind, at the center of our existence, we can seek the Soul.

"Who am I?" is a question asked of necessity by a personality. Other things in the universe can answer to "what" or "how", but no physical thing can answer to "who". Identity is a question of personality, of being greater than itself physical; "Who" needs to be asked by a mind conscious, to us a mind human. Each man, even in a still primitive state, his mind not much risen above that of fellow animals, already formed a sense of "who". When he buried his dead rather than let them be ravaged by nature or scavengers, when he painted on cave walls, or carved images in wood and bone, he sought to preserve identity. His burial hid from destruction a body representing a being that had already formed the question. The person probably had a name and perhaps was referred to with affection. They may have loved one another and the passing of one was mourned with grief by the other. A sophisticated, cynical mind might say that this love was but an expression of one's personal, selfish desire and that such grief was but regret for the loss of this self indulgence. Perhaps it is true, but only a personality is capable of it. However, more likely, the person deceased was also capable of this self indulgence and they could mutually enjoy each other in this love. Love may or may not be greater than a selfish desire, but the body was already more than a collection of flesh and bones, to us moderns of atoms and molecules, and was treated with a certain respect even in death. It was a being, a personality, a "who" capable of giving and receiving with affection the being of another. To a person able with identity, to a personality, the body of man represents a being also spiritual.

In time, the who of personality became religion. Early we began to personify our existence with images of our being. The gods of the ancients were powers with human traits, endowed with the same personality traits of nobleness or mischief allowed us lesser mortals. Mountains, woods, streams, meadows all took on characteristics of human personalities; spirits and faeries inhabited even our homes to work good or mischief depending upon how we treated them in return. We were naive in so easily personifying our universe, perhaps, but we found comfort in interacting with physical representatives of personality. We even personified the planets and stars into what we know as astrology. In this vast network of magical forces and beings we found an early framework of universal supports for our individual existence. Our personality, our "I am", was not alone, for it existed within a vast network of beings around us. There is a sense of magic in this kind of existence where everything, including ourselves, is populated by spirits. But it is a magic that can easily be misused. When this wondrous world of the spirit became used for personal ambition, for satisfying a blind selfishness, the world began to change. By calling on spirits to serve our self interests, by denigrating forces of existences into forces of evil and destruction, the magic of life became employed in a worship of sorcery and superstition. Lazy, greedy, ignorant, we had fallen. We strove to influence reality in spiritual ways unclean, to alter the natural path of things as the universe expressed them for us in our everyday, natural existence for the sake of material gain, to enslave. Sorcery became base, self-seeking, cruel. Until the advent of more modern religions, we were lost. With the coming of more gentle teachings of compassion, meditation, and love, the Church of the world, in its many denominational names throughout the peoples of the world, began its attack on such fallen use of the spirit. The world was taught to be cleansed. The Church effaced our past, erased it from our spiritual memory, and once again gave us the freedom to pursue the path of our true being. We are here now. Yet, through all this, the idea of the soul survived.

Now, in our modern world, we are again posed with the question of our identity. We know that in our subconscious is the repository of much that lends support to what is that each of us is. From that vast pool of both the rational and the irrational flows to the surface facets of our personality. If it were not for the conscious mind, these facets would be our personality, but the universe is more complex now. We have the ability to choose, to influence both the rational and the irrational, the material and, in time, the immaterial. The new mind will of necessity learn to manipulate these, they already being a part of our identity. It will be of no surprise to find in these pages ideas that transcend the rational and enter the domain of the irrational. They are our human dreams.

Also, to a large degree, in our modern world we have been cleansed in our minds of the superstitions that past spiritualism prejudiced us with. We no longer see goblins or faeries or fear the evil-eye. Thus we may approach the idea of personality simply, free of debasing beliefs. Instead, we are free to gaze upon a person and see the personality there freely, like on the face of a child, or in the eyes of lovers. Personality, personified creation, the "who" of reality, tends to arise oftentimes in places unsuspected. One would expect to find it in a congenial setting, healthful and nourishing to the soul, yet it can appear where there is misfortune or pain. It may appear on the faces of the destitute, in the hopes of the ill, or in the faith of the oppressed. Where reality is harsh and cruel one would expect brutality, only to be surprised to discover hospitality and friendship. Such is the puzzle of personality, that which we will seek to identify, and that which is the reason we are human in a world of matter. The soul is beautiful, yet it is also powerful. It is what radiates on the bright faces of children, hides behind the worried faces of adults, only to resurface on the creased, kindly faces of the old. Yet, it is what makes people martyrs or heroes or leaders or healers. It is what rises up against clearly overwhelming odds and succeeds. It is that which is gentle and yet terrible, lovely and yet strong. It is courage, kindness, sensitivity, creation. It is also the only thing in this universe of being in matter, of "what" that answers to the question of "who?"

Thus, free, let us seek the soul. Let us take some liberties and allow our mind to paint for us the images it wishes to present. It is human in exactly the way we are human and what will flow from it will be rational and irrational in the same way we are. Will what will present itself be the truth? It will be exactly as we create it. Individual, a thing of creation, our mind is what ties us each to our personal infinity. Let us see what this personal infinity is, created in the way we had been created, a personality capable of a Soul.

Chapter Eighteen

What Is the Energy? On Belief

WHAT IS THE ENERGY that suspends a thing in its own identity? What is the identity that defines a personality in terms of its greater reality? What is that which swells the mind with life and love?

If we think of the energy that daily floods our planet, how the Sun bathes the Earth in its rays, and how this energy is absorbed planet-wide in the many life forms and chemical-electro processes that span the globe; is it any wonder that our small blue-green sphere is thick with life? To consider that only a small portion of the total energy that reaches the planet is reflected back into space or lost through the dark shadow of the other side, think how tremendous is the light absorbed and how this energy is manifested in the many living things. The planet through its multitudes of life, the great wealth of all its living things, greedily drinks in these rays, each life growing and perishing, evolving until one of its specie can look up at the sun and sky and question the meaning of its identity. What is the energy that would fill our sky with its golden-blue light and fill the world with life and man and the mind with thoughts and beliefs?

Through the milleniums in man's quest for an understanding of nature, it was natural to turn to belief as a means of understanding what were mysteries to us. To believe, as in magic, is to seek understanding through similarity or allegory or parallels. It is to see through an understanding that is more oblique than rational and direct. To believe is as if one were to see something by looking past it or through it and thus taking in the whole and seeing it in terms of that whole. It is to understand through stresses, for it is to put oneself at risk, suspended between the known and the unknown, or even unknowable. Belief is associated with mystery, to understand with the use of a new energy, to discover as an act of faith where error will extol its own price in terms of our being. If we are wrong, we may perish. Yet, how exhilarating it is to seek to believe.

When we seek through belief we raise up an instinctive sense of danger, of risk and of failure. This fear is also a vitalizing force, for it lends energy to our need for survival, to dare even with passion and adventure. It is stimulating to dare, to believe, to tempt existence with a bold act of defiance for the unknown in our abandonment of the secure and the known. How bold to place ourselves at the center of our universe and demand to be recognized! Yet, this is the demand of the ego that dares us to believe. This is how we place ourselves at the center of our identity's existence and seek to commune with our greater reality. To do as we believe we must be willing to be as we believe; otherwise we are but a shadow of our real selves, unable to truly occupy in space and time that identity that is our reality. To be at once one with the universe, we must risk to believe. Then, with that oblique kind of understanding, we are positioned to see in terms of everything else, to see wholistically. Thus we may liberate a new power of understanding, a new energy of our human belief.

This is the vitality of belief; it energizes because it transcends logic; it urges one to follow this belief until it yields to discovery and relieves that stress that had been built up until that final reward. It is a creative rather than a passive state of mind, though it may be contemplative rather than active. It is also what connects us to some viality that urges one to go on even when the risk of failure is great, to have faith when all seems hopeless. It is what energizes that other side of the mind, that level of the irrational that seeks to understand on its own terms. When we believe, we become energized.

We easily understand our universe in terms of the three dimensional and thus are ill at ease in seeking to understand it in a multidimensional way, as in evoking belief. However, it will be our attempt here to show how this belief works in our three dimensional reality and to understand it there. Same as we had earlier been able to conceive of human identity in terms of space and time, so we should be able to understand human spirituality in terms of our identity in space and time. How we do will reflect how we are; how we are will reflect how we are in our spiritual reality, our greater being. With that understanding we will seek to conceive of a universe where the interrelationships of reality are not physical or mathematical or metaphysical, but transcend them to a more spiritual form of energy. We may not yet know how to understand this, but we can begin by seeking through belief as a guide.

"I believe" is an expression of knowledge within the unknown. The statement is made with passion, with force and righteousness, of being as one with fortune and fate. It is in the deeds of heroes, the self-denial of saints, the quest of visionaries. Consciously we can say "I know" or "I am", but it is supra-consciously that we say "I believe". Once said sincerely, unashamedly, courageously, forgivingly, once it is said also intelligently, then we can enter into that domain where we can "know" because we "are" in communion with our greater being. By being at the center of our belief, by acting upon this being and thus placing ourselves at the center of our actions, we are at the center of that in the universal order that has given the power to believe. In "I believe" is the beginning of our pursuit of our greater being.

If belief is a metaphysical state, as being in the mind is a metaphysical state of being, then what are the mechanics responsible for the energy released by belief? Is this energy in the form of a power of telepathy? Or is in the power of a life-after-death spirituality? Or is this energy akin to the life forces that surround our planet and tied to our universe? When we occupy our identity in terms of space-time, are we at once somehow based within this vast network of transcendental forces? Or is all this vitality best expressed as an energy of passion and affection and of love? What are the manifestations of a human being calling forth the power of the universe in the mind's expressions of belief and love?

There is a connection here that should be coming into focus: We are in the mind when we are free to be as we believe; but we are in our personality as we believe in terms of our being.

The mind believes and the universe places it in terms of its being as this belief reflects the greater reality that defines that mind's personality. The person believes as is in the nature of his or her personality to believe; the personality exists as is in the nature of his or her identity to exist; by being in the mind that identity and personality approximate one another more closely, as is in the nature of an infinity defined by interrelationship. We believe, we act, we are, and thus we become more closely associated with our identity. When this association is correct, that is, when it is correct in terms of the infinity interrelationships that define it as an identity, then our belief is correct in terms of our personality identity. It is what is right for us, though we may not know it in our mind; it is what is right in terms of our total being. Consequently, once this freedom of belief is established, as in the freedom of occupying one's space in time, we more correctly occupy in our minds that belief that is our personality. This is so not because our belief is more correct than before; we may still be in error in thinking without being aware of this error; it is because by being free in our belief, we are in our reality more closey as our being defines our personality. The physical pressure of existence on our personality forces us to believe correctly, when we are free to believe. Then the existence that defines our identity and the mind that is our personality are more closely aligned with one another.

Assuming that this freedom of belief exists, even if this freedom is not recognized socially but grasped at through a willed and stubborn refusal of surrender to an accepted disbelief in belief, let us assume that we are free to be at the center of our existence. What now acts upon our existence is what is more closely associated with what it was that had brought us to that moment. The description that serves us when someone says of us "he has personality" or that our personality is in a certain way, or in the description of our physical traits, or our environmental and personal circumstances; these all describe our life as the pressure of physical existence has produced them in relation to our personal identity. The universe had worked on us from the beginning of creation until now to become as we are and where we are. The infinitesimal influences of interrelationship, ad infinitum, never spared us with a gap that would have disconnected us from our personal identity's development. We had no choice hitherto regarding our existence; it was with the awakening of a conscious mind that choice was first introduced in this existence; we were worked on by reality until the time of our consciousness and only then came the ability to willfully influence our future. Now that we are human with conscious minds, by choosing, we can alter the course of our future development and evolution. The burden has shifted and, though reality has not relinquished on us its hold, we are now invited as participants. Each one of us individually and personally can now partake in the future development of our specie, how we will. However, how we will is a contingency that will become increasingly evident in how and what we believe.

From the center of our existence, how we choose, we broadcast our being back into the infinite network of interrelationship in our universe. The more closely we occupy that center of existence as it is described at infinity in terms of our personality, the more closely does our physical existence resemble those characteristics that are our personality. We are, in effect, how we wish to be in terms of our personality. This being in terms of personality broadcasts back into reality our free association within our existence and we become moved and can move more closely in a way related to the being of our personality. We must say "more closely" or "tending to" because there are no absolutes other than infinite interrelationship. Even there the absolute is broken by the ability of growth. To achieve perfection, at this point of our existence, necessitates that we tend towards or seek to approximate, because that is all the mind is capable of. If it were not so, there would be no need for effort and evolution would be an accomplished fact. Because there is still work to be done both at the personal level as well as at the universal level, we must always in our thinking understand that there is no dogma at work here other than the right to seek freedom, for all else is relative to the way it is described at infinity. Thus, how we broadcast our existence allows our identity to "know" us in its universal way and, in return, we can then seek to "know" us in terms of how is manifest our being in this greater identity. Our belief is the energy that broadcasts from within our center of existence in our identity; the energy that is released by this belief is what tends to broadcast from our greater identity to our individual existence. The effect is that our being is directed within its existence by the pressure of events and circumstances as we find them in our immediate environment. The presssure of direction is a physical manifestation of our belief at work in our universal reality. It is the energy that is radiated by our greater being, our greater personality. In effect, it makes us what we are.

This would appear to be a rather roundabout way of saying that we are as we believe, that as our minds believe so are we the persons we are. But such direct simplicity would be deceiving. There is purpose to this exercise: We are what we are only in the relationship that exists between our mind and the physical reality within which it exists. We can say that our mind is actually impotent without its greater reality, for it is against this greater reality that it gains in expression. We cannot exist in a vacuum, we do not have the power to change reality as we wish, and change is always conditional upon whether or not reality accepts any desired action. Whether or not an action is possible is a condition of reality and not of will. Together, therefore, for the possibility of desired change to be realized, as the terms imply, must be accepted by both the mind and by reality.

The same is true for our search of a spiritual reality. We may wish to transcend the physical and experience being that is somehow purer and more sublime. To withdraw from the world and its corporeal cares and seek spirituality through meditation is an example of being forced to recognize that there exists a separate reality that needs to be negated in order to achieve the desired state of meditation. By negating the physical, only the spiritual becomes of consequence and a greater perfection may be attained. However, that may be an illusion and no more desirable than saying that a person's being is as he or she believes without considering why this is true. The monk in his cell, or the hermit in his cave, may have mastered the ability of negating influences of reality to an acceptable level for the attainment of spirituality, but by doing so he is also negating that which had brought him to the point of desiring to achieve spirituality, the greater physical-spiritual reality. Thus, to withdraw from physical influences is a negative behavior, since it does not recognize that it was this reality which made the withdrawal desirable, even if not totally possible. It is as if there were a trap along the road of our development; withdrawal from a total reality, both physical and spiritual, may serve to divert us from that which we are trying to achieve; a greater being. To be as we wish to be, we must allow for being as we are. This being is our greater reality which is both physical as well as spiritual. One cannot be without the other, for they are inseparable. To seek a "short cut" from the physical to the spiritual reality is of no avail for then it is as if saying that a person is as he or she believes irrespective of that person's greater reality. To be without the influences of one's greater reality is a totally egotistical being, since the mind is believed to hold in its inner sanctum the key to all existence. If this is true, then there would be no longer a need for physical reality and existence would be free to negate itself into an ultimate Being. It may be perhaps that because the mind of man has not yet achieved that lauded level of attainment that this total negation of being has not yet occurred. On the other hand, it may be a belief in the ability of the mind to hold such power is in itself egotistical and that the shedding of the ego against such a framework becomes impossible. We cannot negate reality because reality is the definition of what exists. To seek a total ability to negate reality can result only in a negation of our greater reality, which is not the desired state of being. Thus for a belief to believe that it can achieve the object of its belief without the benefit of its physical being may be no more than a spiritual trap, ego diversion. However, such an exercise can have no harm in it, for the mind that enters it only succeeds to negate itself. Its real benefits must lie in a different area, for meditation can serve as a temporary holding area where the mind can re-sort itself in terms of its greater reality as that reality has the time to realign itself with the meditative mind.

Thus: We are in terms of our greater being; our personality is a description of this being in terms of our greater reality; our belief is a description of this personality in terms of our mind. This is the trilogy that makes it possible to be as we consciously wish to be, as we believe. It is from this trilogy that flows the energy of our existence to make us more closely resemble in our personality that which describes for us in universal terms what is our individual, human identity. If our belief is love, then that is what we project. If we believe in compassion and forgiveness, then that is our broadcast into reality. If we believe in helpfulness and charity, in fineness and sincerity, in truthfulness and rightness, in beauty and music, or in humor and joyfulness; then these are all elements of the mind that swell our lives with their special characteristics of personality. And if it is in their negation that we believe, as we will see later in the text, it is to these negations that we will also succumb. We are as we are because of what we believe in terms of reality. If our belief is life giving, full of joy, then our existence is healthful and joyful. And if it is sacred, seeking to merge the heart and mind with the Soul of the Creator, with the ultimate greater reality, then such belief is also a power on which hinges our future development and which brings the reality of Creation into our existence. We are how we believe because by believing we are in how is the energy of our greater being. It is that energy that fills the body with life and personality and the mind with love. It is an energy that can experience life after death, or the magic of Earth forces, or the thrill of minds joined as one. In belief is the energy that to a conscious mind may be called the Soul.

Chapter Nineteen

Images From the Right Side of the Brain

LET US REMEMBER, as if in a dream, as from that other side of the mind shrouded in images of imagination, as if in a game or a dance. Let us seek to see in that oblique way where a direct vision and understanding is impossible but where we know it nevertheless in a state of comfort or as in a recognition. Let us say "I know that!" in that secure way where we are in the midst of a familiar place and a sudden recognition dawns.

We remember virtually nothing of the time of our birth, of when we left our mother's womb, of the time shortly before or shortly after. We only vaguely remember impressions we may have had in later years, or dream of them allegorically as being in a room with a door too low and narrow and wondering "I can't get through that, it's too small!" But a conscious, wakeful memory of the event of our birth is closed. We can know it only abstractly because we know we are all born and because we exist. Hence, we are born and, though we do not remember it intimately as a part of our past, it happened.

At the time of our birth, when the energy that shrouds the planet was released once more into another living being, we entered our body in the present form. We became within this vast existence another living being endowed with organs and functions, a human form that would grow into our adult size, facial and physical features that would identify us, and a sense of being that would ultimately become our consciousness and personality. We do not remember the moment when this took place or how we approached the issue, whether with glee and expectations or with fear and concern, or just accepted being inevitable, a job that needs doing. The material of our limbs, of our hands and hair and eyes, somehow all came together at that time with its own memory of being, of knowing how to be alive, how to breathe and how to eat, how to pump blood in our veins, and how to reproduce. It also remembered how to grow and how to learn and to seek a place in existence, survival. These are memories that already came with us, yet, of them we remember nothing. We also learned to recognize our parents, to recognize our friends and relatives, to feel kinship with fellow human beings, to become individuals in a vast family of humanity. The last lesson we are still struggling with, but it is part of the energy of a planet. From the many different faces of the people of Earth we were released as an individual with our own personality and our own identity. In the event of our birth, we remember.

Our memory is shrouded by our focus on our being. We could have been born an animal or a tree or remained locked for eons in the being of stone, but we were not. We chose our being human, here for but a brief spell to work the work that we need to do and then to quickly vanish from our present form. We are part of the general energy and beauty of Earth, drawn here by her magnificence. Our world is so rich in life, it swimming or clinging or flying in virtually every nook of the planet's surface, even where it offers a rather inhospitable environment. The pressure of life fills the cavernous depths of our man-made cities as well as the depths of caves and oceans. We all cling so desperately to what must be the promise of a tremendous drama to unfold, for we wish to live so desperately. We are focused totally on existence. Yet, there is for some even a modicum of leisure or play, and for us humans there is even the leisure of play in reflection and aspiration and imagination. It is difficult to live; it is work. The pressure of life, of exisence pressing on this sphere, in this corner of the cosmos, in this solidified dimension of reality, has forced us all, plant and animal alike to focus ourselves almost totally on the matter of our being. We are and thus we must be totally. The leisure to remember is rationed stingily only to where the necessity of life can be temporarily suspended, where the demands of the body can be satisfied sufficiently to allow for a small release from its total hold. It is the price of our ego, of being here in the flesh, in a dimension of matter into which we are born.

Our organs know, our body remembers, the skin and fingers and breasts, the color of our eyes, of our skin, the curvature of our lips, of our smile, they all remember for they are alive with existence. When we dive into a sparkling pool and our skin tingles all over, we are remembering. When we look at the beauty of a sunrise or majesty of a sunset, when we watch the sides of mountains bathed in pink light or see the sea catch fire, we are remembering. When we gaze into another's eyes, or at the work of human hands, or hear the melody of a voice; when we smell the bare earth or walk on fresh cut grass beneath bare feet or look up at the heavens, we are remembering. And when we look into ourselves, probe the depths of our soul with meditation or the passion of profound humility, when we raise our hearts to that which has brought all Creation, we are also remembering. We are remembering that we are human in more ways than being focused here, and that we are part of an energy of life that is greater than its focus on this planet's existence. We do not remember that we know these from the inside of our being, but by thinking of them, we are remembering.

We exist at two extremes. At the infinite center of our being is the utmost "I". At the infinite outer limits of existence is the utmost "I am", being. Being is personality and this personality is reflected by "I am". It is present in all things in existence. "I" and "I am" is part of every form that occupies our reality. It is more concentrated in the being of a dog or chimpanzee, most in man; whereas it may be less concentrated in lesser life forms. It may dominate the whale or dolphin, preoccupy lion or elephant, but it nevertheless is present in a fish or salamander; it is present in every living cell or organ or muscle. It may even be in some measure in the insect or tree or stone. The energy that is personality is the focus of existence that is solidified in form as the matter of our universe. It is solidified in what appears to us as inert at the non-living level, and it is materialized as living matter at its greater levels of consciousness. In man, where this energy has achieved its great expression in terms of personality, it is materialized as a being able with consciousness, as "I am". It is in our being, as opposed to the being of other living and non-living things, that this personality energy has the greatest focus. Yet, this is the paradox: Where the non-living and otherwise lesser living things gain their energy of being, their sense of personality, from the utmost "I am", we are conscious at our level of "I". All of existence, stripped of an ego, experiences consciousness from their positioning and existence within the infinite interrelationship of the energy of the universe, where their consciousness of being is more a part of the infinite than of their own individualized cognition; we, on the other hand, are cognizant in terms of the ego, in terms of that independently individualized personality that has manifest itself in "I". The animal and plant and possibly even the inert world are conscious in terms of everything else, whereas we are conscious in terms of our own being within everything else, almost to the exclusion of all else's existence. Perhaps this was our earliest downfall as man when we had stepped from a paradisical support of All that Is into a conscious mind. We ceased to remember; we became ignorant of "I am" in terms of all existence and only now are again groping for it; we became cruel and mean in the eyes of existence. And yet, how wonderful that the energy of utmost consciousness nevertheless supported us and sustained us to the present. We ceased to play in innocence and became serious in our work, in the difficult responsibility we had chosen to take on. We became conscious in a way that few other creatures could understand, became divorced from the order of all existence, and in this separation we earned the ability to choose.

The imagery is clouded here, for we are more focused in the ego than in the interrelationship of our being. We are not part of the natural scheme of things. For example, we refuse to be eaten and we do not have the fur or feathers to protect us from the cold. Even now we dare to do the unthinkable and demand that we be allowed to choose our own space in time. How bold to step from all of reality and to choose! Yet, this is what defines us as human beings as opposed to the being of all other things. It is possible for a dog or a horse to seek freedom; they will escape if caged; but it is not possible for them to state their choice and to stay by it. This is left to the mind of man,to seek the power from the freedom of choice. Perhaps it is also that seeking this freedom, this new energy, that will liberate the other creatures from their difficult struggles with survival. We would not be eaten, but would not other creatures prefer the same? We would seek love and comfort, but would not this same love also be a joy to all other living things? We have a job to do, developed an ego that allows us the freedom to do it, and are positioned to do so. But, in the seeking of choice our focus on reality in terms of "I", the ego, dimmed for us our memory. We had forgotten our mission, in enslaving or hurting others we had forgotten our past; in our preoccupation with the comfort of our own being we had forgotten the future. If we could occupy our own space in time, in the way all other things occupy their own identities, their respective and personal images of personality; if we could be free to choose within our personality, perhaps we could remember what the focus of the ego had us forget.

From a great distance, from the utmost limits of "I am" into the infinitesimal center of "I", we traveled through space and time to occupy our present form. From the vast expanses of time of a painfully slow evolution to the vast reaches of space in the creation of the matter of our existence, we have reached a point in space and time that defines our present being. We are here now. Whatever vast potential was released at the time of Creation, we are that potential realized in the present. What will we do with it? Where is it going? What have we done? We are the present of our universe; we are being.

We are born into the matrix of existence that was ready to receive us at birth. The moment was chosen; the parents we were born to, the time and place, the family and friends to come, circumstances and events of the times, these and all the souls we would meet in our lifetime had already been preordained for us. We chose them. We chose our reality before we were born and with them we chose to travel the brief journey of our existence here. For each moment of time in our existence there would exist a near infinite number of possibilities to respond to our choices while alive here, and for each choice there already existed a near infinite number of responses. It would be difficult to err in such a system of reality, and that is how we preordained our being. These choices were predicated on our individual realities, on what we had done before in our personal successes and failures as personalities in the past. They are also predicated on what it is possible for us to do and on what we choose to do. These ideas of personality need not refer to some great past universe or some long vanished higher civilization; they are the manifestations of daily existence, of humdrum past lives that struggled in the ways they knew how, either conscious or unconscious fortunate or unfortunate. We are engraved in our being with the past. It is on the proportions of our face, in the manner of our speech, in the way we touch and hold things. These have been built up over countless generations of existence to the present. Each separate life is so fantastic in terms of its universe that it is bewildering to think of. Because we are so familiar with existence, we are unimpressed by the miracle of birth and living being; because we are so focused in the ego, we tend to belittle life's greatness. Each life, no matter how mundane, perhaps no more glamorous than the existence of a worm, is in fact great in terms of its total reality. It is tied to an infinity that propels it. It is the realization of the universe's potential as personality manifest then and there. Our ego has become so familiar with these lives as to almost border on contempt, yet that is but the product of its narrow focus. For each human being there exists a vast network of existence, for each deed there is a vast energy at work to bring about that deed, and for each birth there is a vast framework of existences that had already laid the ground for Creation's potential in the making. There is the magic of our birth and existence, for in it is the miracle of a universe's realized potential in a living being's life span. In man the living miracle of being is further heightened by our ego's power of consciousness. How we choose within a life span reflects how we are working to materialize that potential into reality. It is how we remember our past lives. Into this we are born.

We do not remember how we chose our space in time, nor how we were chosen because of our space and time. In past lives we had realized what were our potentials, energized those parts of reality in contact with our being, and then we died when that energy ran out. It was a process repeated innumerable times; it is perpetually repeated by all living things; it was repeated by each one of us to the point of our ego dropping its consciousness of it. We are virtually numbed by our past existences, which is how the ego came into being. In the quiet of eternal re-incarnations we became powerful as "I!" in "I am". In each successive life time we filled the vacuum left by us by our previous existence and occupied that vacuum with our new being. The genetic code, the space-in-time requirements, the adjacent personalities that would be with us were already all programmed into our appearance on Earth. The energy necessary to once again motivate these into existence is one we needed to bring with us. We had to energize ourselves into existence to become born; we channeled the "I" of "I am" into the "am" of "I am"; we energized the ego to become a being conscious of itself. Once so energized, we were able to bring our personality into being in the form we now occupy. This is done so frequently and so easily as to be utterly dismissed by our consciousness. It may not be a perfect fit always, but work implies that perfection is not yet achieved. It may not be that we become who we wish to be, but then the choices we had made in our past lives were our own and their benefits we now reap. We have the choice to make current ones that will change us in a way desired. In this life, how we will choose is how we will be in the future. If we recognize a smile on a stranger's face, or see a personality in another's eyes, they are already familiar to us. We had been before. If an event brings us great joy, or if we are doing what we truly wish to do, or we are where we wish to live or simply be; they are all indicative of our being as it is constructed for us from our memory. We remember them; we are drawn to our past because that is where the present was formed, where the choices of our present being were made. We chose to energize our existence with our birth and, in so choosing, we chose to begin anew where we had left off before. It is because of our present existence that we remember the past, even if this is not remembered consciously. It is written all over us, it is in our personality, it is in the physical characteristics of our being. It is in the "I" of

"I am", as well as in the "am". But in our building of the ego we had divorced the two. To remember, they must be once again reunited: "I am!" Thus we exist within our own creation. We energized our reality; around us is a vast garden of life energized by plant life and animal life and the subtle existence of matter. From the mechanical wonder of insects to the flowing waves of meadow grass shimmering in a field, we are surrounded by the adjacent life spans of what exists around us. To see it, to feel it, to experience it is to put oneself's being into the adjacent being's energy with a special rapport. To lay a hand on a cold stone is to endow that stone with the energy of our existence, it is to excite the stone's atoms to dance with ours, it is to feel the force of its strength. Perhaps that dance is merrier than we know; perhaps the life of matter comes to life only with the energy of another living being. So it is with animals, that we either attract them or repel them. Some even come to live into our homes; we know them. Each thing is born and reborn to begin anew where it had left off when its energy ran out. It is the soul of each thing's existence. It is what remembers our being when we are born. And it is all in memory of how we had chosen.

Chapter Twenty
The Idea All That Is

THE IDEA THAT EVERYTHING is related to everything else is a very simple idea. It harbors within it the
essence of being: That everything is exacty as the pressure of
everything else has allowed it to be. The interrelated totality
of a universe is the final arbiter of reality, since everything, every
event, is as that final arbiter has allowed it to be, as an infinity
of interrelationships, a complete whole of "all that is." However,
this idea of interrelationship which sprang from a most basic
relation between three points, the original idea's basic triad,
has also grown into the most complex idea imaginable: That
everything is related to everything else in All that Is. From this
essence of being, how the universe defines a thing's being,
grows a complexity of being that defies the ability of the mind.
We can never hope to know what the image of an interrelated
infinity looks like, nor can we ever hope to understand all of
its intricacies down to the smallest detail. This level of understanding of the most complex is beyond the domain of idea
but rather is in the domain of belief. The essence of being is
an idea of interrelationship: All that Is is a belief in an idea, in
the essence of being.

We can never hope to grasp directly the significance of All
that Is: it being more in the domain of ideas of the gods than
of mortals; but we can see directly what that interrelated whole
looks like at the level of our reality. It is a characteristic of
interrelationship that the greatest and most complex is responsible for being in the smallest and most simple. The infinite
interrelationship is responsible for the creation of man, an individual human being with a mind. This is an infinite simplifica-

156

tion that is visible to us as human beings. The complex mind of man is capable of individual thoughts; an infinitely complex universe is able to arrange itself into individual units of energy such as atoms and molecules; the greatest, in the manner of everything related to everything else, is forever reduced into its, in effect, inverse: the simplest. So is it with the totality of interrelationship expressed in All that Is.

Through the energy of belief, we are able to encompass the greatest ideas into their simplest values. In turning to the same process that defines reality to itself in terms of itself, we can turn to a belief in All that Is in order to see it As it Is. They are but inverse images of one another, one being the all encompassing and the other being its particular. Each thing is in infinity as that infinity has allowed it to be, and so is it with an individual idea; it is especially true of a belief; once in existence, because of All that Is, they have their own energy.

To believe in All that Is as the essence of being is to be reduced to a basic value of reality in one's space in time, in terms of how that space in time is defined in terms of All that Is. To believe in this idea is not as if to believe in God. God would be more of a belief in an interrelated whole of personalities together as an infinite Totality of Personality. Rather, an interrelationship of personalities still beyond the scope of this exploration, it is more an interrelated whole of things and events that comprise reality which together ultimately lead to the creation of the value we would call a personality. Belief here is still in the realm of idea, whereas belief in God would transcend idea completely and be entirely in the domain of belief, of personality experience. Thus, All that Is is not a religious idea but rather a metaphysical one that will help us define certain values of reality. When we believe in All that Is, therefore, we are not turning to a mystical idea; instead we are placing ourselves at the metaphysical center of the energy that is released by our individual, personal identity. The universal reality of All that Is is infinitely complex but, through our belief, it is reduced to a simple force that can be worked with directly in our everyday experience. It places us in the mind, connects us with the memories of our greater identity, and allows the essence of this identity, this being, into our human existence. Through belief, in All that Is, we become one with ourselves.

In our pursuit of being, we are at the center of our existence. As we are at the infinite value of All that Is that defines the value of our being there, so are we in the value of reality that defines our being here. The greatest is reflected in the simplest and the personality of our being is reflected in an individual being with personality. What we are inside ourselves is what we are outside us. What we see, we hear, we feel, we touch, are all values of reality that have already positioned themselves to be received by us as we reach out to them. Reality, in effect, rushes up to meet us at our every step, as we are in our greater value in All that Is. Our personality exists in All that Is as an identity: I am. It exists in All that Is as a greater identity: I am my being. "All I am in my being is as it is in I am." This is how the two values that comprise our being are the essence of being revealed in the immediate environment of our existence. It is how we touch the world and how the world touches us. As we are here is as we are at our greatest value; as we are there is how reality rushes up to meet us at every moment of time.

We are now posed with an interesting idea: That how we touch the world is how we are at infinity; conversely, how the world touches us is how we are in our personality. The two values of personality, the infinite and the particular, are thus simplified into individual being. Together, they form the rich and multifaceted experience of a human being. The experience of being alive, aware, conscious of the self and of the world, how we touch things, how we mold them, how we create; all become a window into our soul. How we choose or how we seek to explore are now all manifestations of the energy of belief: To believe that the greatest is forever reduced into the simplest, that as we are at infinity is how we are in our being here, in our thoughts, in our moods, and ultimately in our chosen actions. Then, in how we are in our being is how our being is reflected by reality. In the essence of our being, the universe touches us.

If the basic value of our being is personality, then personality is an individuality that has substance in universal terms. We may call personality that phenomenon that has the ability to form a body and give it that special identity that is revealed in its features or beauty or presence. It shows itself in behavior, in our personality traits, our moods, our ways of handling reality and its consequences, generally those characteristics that distinguish one being from another. To do all these, it must be an energy within the universal order. Personality may be no more than the animism of a simple life form, or it may be the complexity and sentience of higher life forms. In man, to our knowledge, it has reached the highest level of sentience with an ability of consciousness. We have the ability to express our feelings verbally, conceptually, as well as artistically. We are able to express ourselves in ways that transcend merely the need for survival and can express compassion or charity or love. Though we tend to swing within a wide spectrum of values from the egocentric seeking of power to selfless sacrifice, from the passionate and lustful to the gentle and serene, we are able to moderate these extremes with our ability of consciously choosing our moods and behaviors, even if not perfectly. Thus, we can define ourselves in terms of our consciousness; whether we are lustful or serene depends upon how we are able to work with these values of personality and control them according to our personal beliefs. If the force behind existence is the raw energy behind personality, the energy behind the soul is more subtle with a finer ability to choose personal existence in All that Is. Our ability to choose, and the effects of those choices, are the windows into the soul of man, though they are nothing more than what materializes in terms of the greatest and most complex. Personality is the simplification of All that Is in an individual living being. Personality consciousness is the simplification or particularization of a greater reality in man. Ultimately, personality as a soul is the greatest reality simplified in an individual being as that person chooses within the universal order. How it chooses is how reality will rush up to meet it, or, how it chooses will define that person's being in All that Is as it is. The ultimate simplification is the human soul.

The energy of personality is thus a basic value in the essence of being. It is a real value because it materializes itself as the living body of man; it is potential value because it is capable of producing a soul. Yet, it is the raw energy of being that permeates all things. It is evident in all living creatures, as much in the whale as the horse, the mouse as a bird. It is in fish, trees, flowers, fruit, as it is in bees, or frogs or lizard. It is somewhat evident in the very primitive life proteins same as it is evident in the more advanced creatures in the company of men. Within the framework of these ideas, the personality is reduced even further into any thing in reality that has an individual being. All things are defined in terms of All that Is wherever a part is defined in terms of its whole. There is energy of personality in a brook or hill or a forest lane. When the light filters down into a clearing or when the waves reflect the silvery light of dusk, there is the energy of being there. When we touch a stone or build a temple, when we tend a garden or mend a wing, they are the energy of being revealed in their own way. These forces may be evident in mountains or rivers or great works that have survived time but built by the hand of man. Perhaps this is what the ancients had tried to capture in the great temples whose being had survived to the present. They were capturing the energy of being, of personality, in works that would endure through all time. There is magic to that kind of belief; it survived to the present. That same energy may be as evident in the grass on the field or in the sand on the beach, forever changing and yet able to endure. The energy of personality flows even more with interaction. It magnifies in the presence of a gathering of like mind and like belief. Where it is tended with love and affection, with the strength of conviction and the power of personal sacrifice, it grows still fuller into the consciousness of a soul. With grace and tenderness, with serenity and compassion, we awaken in us the basic value of our being.

In this is the miracle of All that Is. It is an infinitely complex simplicity expressed in an individual being as a body with a mind capable of producing a soul. It is capable of love. It is capable of gentleness and charity. How sublime this existence! The more a living being is capable of affection, the greater is its capacity for the energy that is personality. The more it is capable of personality, if it so chooses, the more is it capable of a soul. A soul is a conscious personality. It endures, it survives eons, it is the value of man in All that Is.

How fortunate for those for whom this pursuit of a greater being in All that Is is natural. For most it is work, yet how miraculous that is at all achievable. There is not always the desire in us for goodness, for it is a desire which we must learn in life. To reach for goodness is an arduous achievement in itself; to succeed at it is to fly in the face of all that seems reasonable. Is it not easier to simply lie or to steal? Does not a child first learn to self seek with untruth before it can discern that such behavior is unacceptable to its superiors? So is it with the natural personality of man. We tend to first grab at what we feel should be ours. It is only later that we begin to discern that such behavior is not desirable in the advancement of personality, in the formation of the soul. When we first touched things, we crushed them or tore them mindlessly; it was not until later that we learned to hold them without damage; it was not until we became more advanced as a mind with a soul that we learned we can create or build. It is the same with the development of personality as a value of being. At first we err, then we build and create and enliven.

The mind naturally wrestles with the conditions of its physical existence. Either it is aggressive, egotistical, self seeking, or driven with the seeking of power; or it is submissive, careful, and selfless in a manner more suited to finding agreement. A personality need not encourage trespass, it should discourage it, but nor does it seek to trespass on or overpower others. A mind can be strong by the virtue of its convictions, yet docile before its personal reality. The elements that confront it at every moment of time are the real values of its being in All that Is, they are its creations, but they are not in its confrontation. Such a mind learns to work with its reality, is fearless in the world of its existence, and gains strength from the elements of its personal reality. The universe is energized by our being, as we are energized by its being. In time, we learn that it is more advantageous to give of this energy than it is to take from it. In such a mind are the beginnings of the seed from which may be born a soul. It is patient, and it believes in the condition of its existence.

We are all personalities, whether in this existence or in that existence of which we but dimly perceive but from which flows all the energy of new life; we are all traveling together on the surface of this planet, at this time, within the elements of this existence. For each one of us exists a vast phenomenon of existence, a panorama of reality that totally envelops and defines our personal being; within this vast being is formed the identity that is our personal being, our personality. From these two is forged the individual existence within which we live. We touch the world, we touch the personalities of others, and we touch the existence of new living beings. Our children are personalities that have come into our being; we are the personalities through whom they found physical expression. How we touch these is how we project our personality into All that Is, and how they touch us is how the universe reveals itself to us. In the manner of Infinity, it is all taking place on a finite value, a sphere, where interraction between souls is a virtual necessity.

Thus we choose in how we are. We choose our friendships, we choose our environment, and we choose how we will affect the world. We either do this through agreement and care, or we do it through force and brutishness. In the extreme, we do it with violence and death. These all reflect on our personality. In the simplicity of existence, they become fixed on our identity and we will carry them with us through the eons as a mark of our being. We will then either be personalities conscious or unconscious, either capable of the soul or merely animism. How we believe, how we interract with one another, how we exist, how we create, how we energize our reality and how we are energized by it is how we are as real values in All that Is. When we become conscious of this, we take on a new meaning in the universal being, in All that Is, and we partake in the idea of the most complex reduced to the most simple: We take on a soul.

"It is good."

Chapter Twenty-One

Patterns of Triads

PATTERNS UNFOLD from our basic triads: Belief, Being and Idea become Energy and Remembering and Identity. All that Is unrolls into an interrelated whole, a universe of Life and Consciousness, elements of a Soul. So much of what we are aware of is superseded by that of which were unaware: of love, of will, of compassion; of a reason beyond reason where cells multiply, the heart beats, dreams are born. The mind had long forgotten the primordial drives and desires that had evolved our species, populated this planet, and brought to us the passion of the reasoning and unreasoning mind. These are all forces of life, of the spirit of human beings, of the cohesion of community. We survive, we struggle and suffer, but we also marvel at the stars, awe at the miracle of life, and revel in a sunset or a sunrise. In the energy of tall mountains or rolling ocean waves is locked in the force of delight and beauty and the freedom of being alive. We are in the body, born of man, and pressed into the network of existence in a universe that had already defined for us our being. This is the magic of the forces of the triads of interrelationship: Three is formed into an infinity, and from infinity through Being is deduced one: I am.

It is through our being that we live. It is through the personality that is evident identity. It is as a definition from All that Is that we understand this identity as personality. We occupy our own space in time and in this defined being we occupy the identity image in the universal order that defines our individual existence. Thus, our being learns to become conscious in it, our personality learns to become one with it, and we work with it as it works with us. But this knowledge is also gained in

another way, unthinkingly, non-reasoned but experienced in
the way one experiences a dance, or spoken as one would
in a song or poem. It is in this other way, this oblique experience
of being, that is defined that other world of the spirit, of emo-
tion, of good and evil, of living and dying. That other knowledge
flows through us like water, or grains of sand, stirred yet undisturbed. In being a personality an identity is an ancient
knowledge that dates back to the beginnings of a universe, the
first stirrings of life, the mystery that had been celebrated for
eons by our ancestors in their worships and songs and superstitions. We need no longer worship in primitive ways, though
we can seek the soul. We can seek it with our modern minds,
being in the mind, the spirit liberated, simply and sincerely,
as free beings. Reality is then an ally, if we seek it that way.
It becomes all a function of how we see reality, how the personality had chosen its given reality, how we became human.
How these are chosen is how we define our greater being and
is how we then materialize our reality around this being. We
live in terms of how we are.

Within this contemplation is a universe working with us. As
our consciousness of our personal space in time grows, so does
the awareness of the space in time identity of each thing within
our personal reality. It is as if things have a life of their own
within our space in time reality, in relation to our being. Each
thing, each event, each person is a phenomenon defined by
its interrelationship within the network of All that Is. The greatest
is defined in the subtlest; the smallest is a function of the
greatest. It is an energy, a vitality that flows from the definition
of this interrelationship to energize and render a state of being for each thing or function. An event is because of everything
that is around it. A thing is because of how is everything else
to infinity. Its image is already defined at the outer limits of existence and from this infinite definition is its definition in space
and time, here and now. How much more fantastic can the
universe be when understood this way? Can it be understood
to be capable of creating a being so conscious, so alive and
possessed with an understanding of this livingness, even if
understod only in that unreasoning way, that this being could
become defined in spiritual terms? Can this focus on all that

exists be a focus that, when worked back through all the limitless possibilities of infinity, become redefined as a being capable of life and a soul? Is the personality that shines through a human being an image of that soul? There is a kinship between ourselves and our reality. It is best defined as a focus of all existence in the vitality of an individual living being conscious of being in reality. The universe works with us there.

When we walk the Earth, we are walking on it not only with the soles of our feet but also with the visions of our mind. Thus we are one with our reality and participate in its creation. It is from the other side of our mind that we energize our being as we consciously accept that reality with our mind. What our mind says to itself is also what the unreasoning mind says to itself in its real definitions. To say to another: "God Bless You" or "I Love You", has deep and significant interrelations. We are truly bestowing good will or love on another in ways the rational mind cannot comprehend. But we do it naturally, anciently, and with sincere conviction. If the mind responds what to it is great beauty or enchantedly responds to a sight or sound, it is exciting these with the power of its energized being. The object of this attention by the mind responds to us in its own way, perhaps in ways so subtle that we are as yet unable to detect it. It is nevertheless detected by the focus of All that Is. If we caress our cat and we elicit in it a sense of affection shown by purring, that affection is also in its totality personality. In a human being emotions are perhaps even more developed, more intense at the personality level as well as at that level that defines personality. But how much more intense is it at that level that defines all Being in All that Is. Think of the soul being created there, a focus of billions of affections expressed in all the ways that living things are capable of expressing it. Now work all this back to the mind gazing upon an object of admiration or beauty. The energy is there as billions of interrelationships are redefining it in terms of how the mind sees it there and then. It is an energy of affections, of ideas, of virtues and sacrifice, of will and of compassion; all of these are elements of the soul and they are present in where steps the foot. Each atom realigns itself in terms of our being, in relation to All that Is, as we pass it with our presence. How we pass it in our minds is then the image of how we are. How we are in this being is also how we are in our soul.

There is magic where our feet touch the ground, where the eyes gaze, where the mind perceives. From all the memories of existence to that moment in time there had been definitions that had realigned themselves until the moment chosen in our steps. All the pressures of existence that materialized our personality became defined in that moment when our being touched the Earth and the Earth responded. This finite act of being, deep from within the recesses of physical space and from the dawns of time is remembered and energized by all of Creation. Singly, it may appear to be such a small and insignificant act. Yet, how powerful are the forces that are defining it in All that Is? These forces of being surround us at every living moment of our existence. We are surrounded at all times by the magic of our creations.

The magic described here is one of fineness and supreme subtlety. It is in the principle of the greatest defined as the smallest, the most dynamic in terms of the most serene. Being is a power all its own, bestowed upon it by the nature of existence in our universe. It is alive, it is sensitive, and it is a force of energy that realigns all things in terms of everything else. It is in the energy of belief, for it is truly visible only there. At our normal cognitive level, these forces are not evident to us. We do not consciously know that the universe realigns itself in the process of every step we take as we walk the Earth. This realignment is in the greatest possible subtlety, redefining itself instantly in the presence of that single step. Nothing seems gained; to the eye and mind there is only the obvious awareness that the foot is in a new location. But at the level of a more serene contemplation, the stage had been set for the foot's next step, for the redefinition of its next encounter, for the energy of the presence of the moment to reenergize all the things that will be in contact with that next step. The greatest force of the universe is evident there in only the most subtle way. Through contemplation, the mind has a minute access to it. The soul had shifted but by the slightest degree, but the powers of the universe had shifted themselves at the greatest dimensions. The surface of the ground walked upon is felt by the soles of the feet and but the slightest sensation of energy tingles and reaches the brain. The mind, in its immense and unreasoning way, is deeply affected.

There is immense repercussion in all we do. The personality is deeply imprinted with the power of all being and thus it acts within reality. From reality flows the power of its being back into personality and thus it responds to its new environment. The personality is altered and the process repeats itself. Now reality is altered and all of being is imprinted with personality. There is a constant exchange taking place between the being of personality and the being of reality. From between them are created the patterns of everyday existence we would call life. They mutually create one another naturally, though this creation is often too subtle for the awareness of the reasoning mind. We are more aware of it at the level of the unreasoning mind, and totally aware where the unreasoning mind interacts with the patterns of reality. We may wish to take a walk. But think of all the other things that we could have done at that moment in time, each with its own results, its own interrelationships in space, with its new imprint on personality and reality. The personality-reality complex emerges as having a separate being of its own. From the many patterns of being, from the trilogy of the energy of belief, the memory of being, and the idea of identity, there flows a new definition, a new force of creation. It is the personification of reality, the energizing of physical being by the presence of the soul of man, by the personified mind and its many multifaceted images of being. We personify reality by the very virtue of our being in it. But this is still an unreasoned, unconscious being. With the formation of the soul, the universe sings when we sing, glows when we glow, and rejoices with affection when we do. To be loved within reality we must be able to love. There is a power here that is difficult to describe, being more in the domain of the unreasoning mind than the rational. It is more in the domain of that which describes our personality, yet moves reality. As we do, the universe does with us; as we feel, the world feels with us.

We can probe this in a metaphysical way, but it must be sensed in that oblique manner of being or belief. The atoms and molecules we displace with our being are filled with a sense of our being in that displacement. They have an energy of their own and we merely activate it. In our mind is the energy that can make atoms dance and worlds move throughout the cosmos. This is the great reason of unreason. It is also the great power that can have a universe project for us the matter of our individual definition as a personality. Then, we are born into this human being. That is the only way we can understand it, for then we experience it as man.

We must take this line of unreasoning reasoning still further one step. We cannot project ourselves rationally into this world, being still ignorant of the interrelationship mechanics that make it possible. We cannot know, yet, how our personality affects reality nor how events desired by the conscious mind can be made real. But we can position ourselves within this state of events, work with them and manipulate reality in ways that will achieve for us our desired goals; we can further work on events by how we position our personality in reality and project our being into that reality. How we are, how we do things, how we touch the world around us with our bodies and our minds are direct communications into the reality being manifest by our personality-reality complex. Thus, it is in our personality's interest that this contact with reality, how we are positioned within it with our minds, is also of some awareness to us. Then, we become less ignorant of the consequences of our being, and we become more conscious of how we are. How we approach our existence reflects this new awareness and the environment of our individual existence begins to better reflect this consciousness. Reality becomes more familiar, more a reflection of our personality, and more energized by our being as in the mind. Intent that is beneficial creates its own results, whereas energy that is malevolent will equally reflect back into that mind's reality. These are subtle influences generated from deep within the mind, but they have the power to move a universe. So released, they may work on us for a great period of time. Our soul is modified by these forces. But in their more subtle forms, to our reasoning minds, they show up in our everyday world. In effect, our state of being in the mind is a powerful effect within the domain of the unreasoning mind, in the domain of the personality-reality complex of our existence, though it is a weak effect in the domain of the rational-conscious mind. This is a force of which we must become conscious if it is our desire to become one with the mind. That is how we can position ourselves in our soul's identity within our reality. There is great energy there.

So we have brought the patterns of our basic triads down to a single force that we may use to help us understand the soul. That force is the energy of a personified universe, a reality endowed with the mind's energy of our individual being. The personality occupies reality and in so doing is also energizing it to conform to its singular identity. It does so by moving all those other forces of infinity that define the universe's reality. If this is so, then personality is an ancient and powerful value within existence, dating back to the deep beginnings of life and mind. Developed in us as an individual being, human, with a sense of identity, "I am" and aware of its existence; we may now seek to work with it in ways that can help us seek our greater being. It is what can help us become more human, more in tune with the creative forces of the universe we live in. We create reality with our being. We do not know how we do this, being more in the domain of the unreasoning mind than the conscious one. But it is a real force, the force of being in the mind, when we occupy in space and time the personality-reality that is our identity at infinity. Locked within these powers are the definitions of the soul of man. It is what makes us creative or hopeful or helpful or joyful. We are human beings capable of a soul. We can gain this soul by working with the mind to personify our reality in the image of our being.

Chapter Twenty-Two

When We Are in the Mind

WHEN WE ARE IN THE MIND, we are at the center of our total exisence. When we seek this being in the mind with our will, we are realigning all the forces of our existence that define our being. When we are at the center of our existence we become, in the personality forces of our being, the true value we occupy at infinity, our greater being. In the reaching for this being, our personality realigns itself with the forces of our existence. Thus we realign our will in the image of our greater being. This is the formula of the soul.

We can reach out into the universe and seek the soul at the many levels attempted throughout time. We could seek it in religion, in meditation, reincarnation, in mysticism; but it is at the level of existence, of being, that it is mostly manifest. We are. We are our personality. In our personality is the distillation of that value that is our soul. In our existence, the soul is manifest in the phenomenon of our physical being. The soul, our greater being, is most visible in the values or conditions or mannerisms of our individual being.

If we are reincarnated through time, then this being is of necessity modified by how we had been in the past. We are not born anew in each incarnation but instead are reformed within the definitions of being we had left behind in our past life. We are shadowed by our past existences. We cannot escape this, for it is the value that creates for us both our personal reality and our personality. It is what is for us our present being in this lifetime.

Thus, our definition at infinity manifests for us our existence. This definition is being continuously modified by our state of being in it, by how we do things, how we think, how we feel, and how we choose. It is also modified by the state of our personality, the kind of person we are. Finally, it is modified by the state of being we had left behind in our past lives. How we were in reality in former incarnations redefined our greater being at infinity. This greater being in turn rematerialized for us our world when we were born into the time and space of our current existence. Because our greater being is what distills itself into that value of our individual existence that is our personality, we continually reflect how we are at infinity. How we are there is then in turn how we become here. It is like a cosmic shadow that we cannot escape, but which can help us with its guidance into our future. The kind of person we are is determined by the many progressions of our past lives; the kind of existence we have is determined by how we had chosen those lives. Now, we can choose our existence by being the person we will.

We create our reality with what is for us the unreasoning mind. By its definition, the unreasoning mind is beyond reason. It is in the world of mystery, the irrational, dreams, or in the world of mystical spiritualistic revelation. The unreasoning mind is that force that is a product of infinite interrelationship, the force that holds together atoms as well as that which forms for us our physical being. It is responsible for the division of cells, the color of our eyes, how we look, how we laugh, how we project our inner personality on the outward features of our being. We do not penetrate the unreasoning mind with our conscious or rational mind for it is beyond logic. It is more like belief, the mind as seen from infinity, as directed by the state of being of how is everything else. In essence, we create our reality with the unreasoning mind from that value at infinity that represents for us our identity. If this identity is then expressed in our personality, then we create for ourselves our world, through the unreasoning mind, from how we are in our personality. This is a rather roundabout way of saying simply that we are in reality as we are in our personality, but it is an exercise that is necessary to show how we create our reality.

In the unreasoning mind is the essence of our being. It is the unreasoning mind that controls what things will be, their color, shape, place; it is the personality that controls where our individual being will be within these things. All of existence is the product of the unreasoning mind; the rational being of a personality is the product of the mind of our individual existence. Between the two, between the unreasoning mind and the personality, modified by the state of being of past lives, is forged the being we individually occupy today. Where are we? How are we? Are we what we will? These are all answered by the unreasoning mind in terms of how our personality is within it. To change these requires that we then modify that over which our conscious mind has a degree of control. We can control our state of being through conscious choices regarding our existence, where we are, what we are doing, with whom. But we can also effect change by choosing to modify elements of our personality to more correctly reflect the person we wish to be. Again, in a rather roundabout way, both can succeed in creating the same effect. But where one is useful because of its immediate effect by physically creating change in response to our choices, the other is more interrelated into the state of being of all things and hence more intense, since it is affected by the state of being of everything else. One is in the domain of the rational, through direct action; the other is in the domain of the irrational or unreasoning, through the indirect action of the state of being of everything else as directed by being of personality within one's greater identity. Both always work together regardless, but for the purpose of illustration they are separated here. Whether we focus on direct action or on personality, we are working with a universe working with us.

Thus, if we choose, we can elevate our mind to the level of a universal consciousness. By occupying our own spase in time, we occupy the state of being of our individual reality, our identity. By being conscious of our personality in this reality, we are able to glimpse into the value that we are in this identity. By choosing to be conscious of how we are in our identity, we can then become in reality how we are in our personality. This reality is created not by the process of rational thought of which we are conscious in our minds, but by the process of the unreasoning mind which is conscious at infinity. In times gone by, this would have been explained in religious terms of God or holy and supernatural powers, but in current times this idea can easily be understood merely as expressed in metaphysical terms. We occupy our own space in time and are conscious of it; we are conscious of ourselves in it and choose our personality. The mystery remains that we are born into this world and that we die from it. It is a mysterious process, but one which is demystified by the fact that we are. We exist in our being and thus the unreasoning mind is creating for us our reality as we are choosing it in our unreasoning mind, in our image. If this image is a mirror of how is chosen existence in the universe, then it is only a function of how is put together the universe. What is of concern to us here, however, is not the image of man in relation to the image of God, but rather that we are each instrumental in how we create for ourselves our reality. We choose our reality in terms of how we choose our personality. When we are conscious of this, we begin to create our being.

In this order of things, in the manner of the unreasoning mind, we are the masters of our reality. We are born where we are to be; we are in our personality the person we are in our greater being; we are materializing for ourselves the circumstances of our being in terms of how we are as an identity. If we are unconscious of this, then the unreasoning mind merely continues to perform these functions without our awareness of it, as it had been doing for millions of years. If we become conscious of this, we can begin to exert a conscious influence on how the unreasoning mind is creating for us our reality. We can begin to choose actions relative to our personality that are consistent with our new consciousness. When we begin to restore where we had destroyed, build where we had made decay, brighten where it had become drab, then we are beginning to exercise this new found consciousness. We are now entering the realm of the soul.

Because we create our own reality, when conscious of it, we create a reality in the image of our soul. This soul is evident in the eyes, in the work of the hands, in the melody of the voice, in the thoughts and care that go into our every action. It shines through everything we do as well as in our hopes and loves and aspirations. It is the image of our greater identity focused into this reality and materialized here as our person with personality. The personality that can look back upon this and be conscious of it is a person that can look into its soul. It is a mind and being that is sensitive of its feelings as well as sensitive of the feelings of is fellow beings. If the soul is graceful or serene, skillful with beauty, then it is our newfound consciousness shining through in the skill of an artist or the works of craftsmen. We will find it in the words of statesmen and philosophers, in the cares of healers and friends. With warmth and compassion that can tame anger and fury, the soul gradually creates its own reality in the image of our conscious identity. It is the strength of the unreasoning mind bent by the force of a conscious soul to create our reality in our image. It is the power of the still and silent over the violent and voluble. There is immense strength in this new serenity for it is the strength of being over chaos. We have it when we have the soul.

Perhaps we have been unconscious as a specie or race for so long that we had forgotten the power of being in the mind. Untrespassed, free, masters of our destiny through the art of agreement over coercion, we have the power of a soul over the chaos of the unreasoning mind. It is the power of a wilderness taming itself. It is the power of love and compassion as well as that of strength of being. From the birth of the universe to the present, we had been creating our reality with the unreasoning mind and had grown powerful within it. We have the mind. Now we can take this power of reason and become conscious of ourselves as a consciousness and thus have the soul. The past is etched on this soul; all of our deeds through time had been interrelated into that value at infinity that is our identity. We are now faced with the stark reality that we are who we had been through time. In the here and now, in the body we occupy with our being, is the depository of all that we had been and that we had done. It is like a shadow that presses upon us from infinity in all that we do and all that is done around us. We encounter this shadow at every moment of our existence; we are forever subject to it. With the soul, conscious of our being conscious in the mind, we can begin to identify this past and begin to work with it to redeem the wrongs that had been left in our path. If we had been unconscious in our past, we can be forgiven, but conscious now we must rebuild that which we had destroyed. It may be that the destruction of our soul is light, but it needs tending regardless. Like a garden gone wild, it is up to us to cultivate the beauty of our new being.

We are followed by the eons of past lives and yet we cannot remember them. But they are here with us on our visage and in our hands. In our courage as well as in our dreams are worlds that had preceeded us throughout the vast waves of human existence. We are but inheriting this, but it is inherited by each and every one of us in our individual being. When we are in the mind, we are the center of this inheritance. When we are in the soul, this inheritance is in our individual being. We cannot remember them but, through belief, we have the power to bring out within our personality those traits of ourselves we wish to cultivate and those we wish to leave behind. We need not suppress them but rather should face them squarely and accept or reject them in a conscious and understanding way. We do not need brutality or theft or deceit or malice. These are vestiges of an unconscious past that is alien to the soul. To be in the soul, that which destroys the personality as well as that which destroys another's being which trespasses on their identity, must be superseded by that which has the power to forgive and to heal. Then we are realigning those forces that define us as a personality. When we can see the force of charity or of compassion, we are realigning the eons of past lives. When we seek what is beautiful in another and trust our fellow being, we are reaching back into time for all those times our being was mistrustful and had been mistrusted. If we had wronged in the past, in our personality is the power to undo those wrongs. If we had tortured or slain, then it is in us to forgive and to comfort where there is pain. If these are not done, then we merely repeat the past. To be born anew into the next life as a conscious soul, it is necessary to not repeat the past. To do rather than to undo is the motivation of as being with a soul.

Our existence is an encounter of personalities and of being. Our mind identifies and defines that which we encounter in our reality, but our soul identifies and responds in its own spiritual way to that which we encounter in our being. We are human at many levels of existence, but we are just rising above the first level. We are first conscious; we are in the mind. Then we are conscious of this consciousness; we become in the soul. Finally, we create reality in our image. Each of these is inherent in the other and all are exercised in our daily lives to some degree. To rise to that degree where the universe works with us is an achievement suitable to future man. To rise to this level requires that first we do not trespass against one another, that we seek agreement, and second that we step beyond and deal with our fellow beings in a manner of compassion. Then we will interact with the multitude of beings that we encounter in our daily existence in ways that reveal for us our soul. How we interact with them, the persons we meet, the anger or kindness that is encountered by us are all indicative of how we are as a soul. How we had been had made us how we are. We have only the future to change. Through how we treat ourselves and how we treat our fellow souls in this world is the path to how we will be in the future of this life as well as the many lives to come. Thus, in the end, we create our own reality.

Chapter Twenty-Three

We Are Conscious

WE ARE CONSCIOUS of the self as personality. We are familiar with our intimate, inward being. How we identify ourselves, how we know others, the presence in their eyes, friendship, are all how we are conscious of the personality. All are images of the personality we call the self. We naturally respond to this self in the ways we know ourselves and others. On a more conscious plane, we know this self in terms of the mechanism of a universal interrelationship, in a way that is conscious of itself. In our self, we are conscious as a mind; in our greater self we are conscious as an infinite identity materialized in this existence as a being possessed of personality. We are conscious of this self, from our greater being, as a soul personalized in the self: "I am."

Thus, we are conscious at two levels: One is the level of consciousness where we perceive ourselves from a personal point of view, the other is the personality as perceived from the point of view of everything else interrelated to infinity defined by our identity. Both define the same personality, but both do so from opposing perspectives: the inner and the outer being. Our consciousness does not yet permit us to understand this fully, but we can attempt to see it in the way that it can see itself in terms of its greater, outer self. We are first as a personality defined by our greater, interrelated being which is our life's essence; and we are second as a personality that has become conscious of the self. The second is familiar and natural to us, whereas the prior is what is still formative in our minds, still in the realm of metaphysical thought. We are not yet conscious directly of the greater self, but we can be conscious of the person we are.

In terms of how we are, how we see ourselves and how we are seen by others, may give us a glimpse of how we are in our outer being.

In our personality is thus an expression of our greater self. It is a projection of our identity into our being in this reality. It is our life's essence, our greater and inner life system that reflects for us our personified existence. Our personality is thus a precious value, for it is our connection with our living and dying, happiness and sorrow, our image as a human being. If it appears we are in the here and now either friendly or fearful, warm or withdrawn, pleasant and generous or guarded; they are all reflections of the personality that defines our being. We are surrounded by it, guided by it, project it from within, and ultimately are led to either live or die by it. There is a force to our being alive that is perhaps the greatest force of our universe. It is the force of personality as our soul.

How spectacular to characterize our most intimate self, as well as our greater identity, in terms of the energy that is expressed as our personality. We project it outwardly as simply as we live and breathe, laugh or cry. It colors our every feature, forms for us our limbs as well as the lines on our face; it even penetrates all we think and do. We are signed by our personality indelibly in all we are as a person. Our identity makes itself known. When we know this, when we know ourselves as a personality, we gain insight into who we are as a greater self. It is that outer, greater self that gives us the being we inhabit in our everyday existence, for that is how is projected for us our present existence. We are who we are in terms of how we are.

We project ourselves into this reality at birth and thus remain surrounded by our greater being's projections throughout our lifetime. It is like being on a track we had chosen but from which we have the power to switch onto other tracks. How and where we live, the manner of our lifestyle, our friends, our community, children, our attachments and possessions, our travels, all project for us the manner of our being. We are also projected in our sensitivity to things, our loves, our amusements, our work. All enliven our being with our greater being. This being is connected by our personality, displayed there and projected both into our present reality as well as into that greater reality of which were but barely conscious. If this energy is manifest in us as personality, it is manifest around us as the state of our being in it. How we choose existence is as much evidence of the soul as how we think or smile. Together, this outer and inner being or personality form our total being within the network of an infinite interrelationship powered by the living force of our universe. How complex and yet how simple! Our greater being is thus focused into this dimension as an individual, singular, living human being as a totally unique self. How cherished is each human life!

We may easily forget the greatness of each human being. We may forget this of all things. When devoid of a soul as the underlying spiritual force of existence, the world becomes drab. But how great when seen that all things are the projection of a greater self into this reality. This projection is the spiritual force of an infinite focused into this reality. All living things on Earth are born, somehow, exist in their living form for some period of time, and ultimately die. How spectacular! They focus here, choose their existence, live, and are released from their chosen form. It is almost as if they descend here from some more rarified existence only to labor here each with their individual assigned task and then return. What marks us is that we have the power to be conscious of this. We can speculate on our life as well as on our ultimate death. We can think in terms of our personality and even think in terms of a greater being. Animals have personality, being in the same mechanism that defines our being here, but they are relatively devoid of consciousness. All respond to some form of affection, have recognition, care for their young, even display play for its own sake. They are probably equally puzzled as we are when tricked or hurt or treated irrationally. To some degree, all even grieve death and are afraid of it. These are universal forces that identify us all as living things. The mechanisms are the same; only we can think about it and choose. We are conscious. Where animal consciousness ends ours begins. We fill the void left behind by this lack of consciousness and are this planet's conscious beings. The burden of this consciousness, this action of choice rather than response, is the burden of our existence here. It is the responsibility of the master; it is the image of our greater identity. Having this responsibility is what makes us great as conscious beings.

There is a vast difference between our existence and that of our fellow animals. Both share in a feeling of life and affection. But only we have a conscious ability of charity and giving. Animals, particularly mammals, may care but seldom if ever give for the sake of giving. They may bring to their master a trophy of their conquest, but that may be more a function of bringing their prize onto their own territory. Within their animal hierarchy and pecking order, animals seldom have the luxury of generosity. One takes what one can. The animal personality accepts gifts but seldom is able to return the gift in kind. Kindness may be returned for kindness. But would kindness be offered first? It would have to be a conscious act. An offering requires a premeditated choice of giving. It would be the act of a conscious soul. But there is a soul in all living things; it is evident in their eyes, their form, their movement. The energy of an infinity is focused there, but in man it is human. We are conscious of ourselves and of all around us. We project this consciously. We give.

When man first became conscious as a being he learned to give and our development became an explosive evolution. Our social structure changed from being a quasi animal hierarchy to that of a more progressive order of law and justice. We could progress beyond mere survival and enter a realm of surplus. We began to do things for the pure joy of doing, to decorate for its own sake, to have ritual, to share in banquets, to hope for the future. When man began to create music, myth, works of art, dreams and religions, he stepped from the animal world into a world of consciousness. To love another and to give became an incredibly elevating act of human evolution. We could give of ourselves, willingly, cheerfully, confidently, and in doing so we were rising above a more animal and primitive past. When we chose selflessness, we became human and the personality changed.

Through many lifetimes, the energy that is our identity changed with the ability of giving. In each successive life form we projected this new self into this reality and had seen in this reality a new beauty, a new refinement. To give without the calculation of a return gift is to elevate oneself beyond pompousness and self gratification. To give with humility, with love, is to pronounce one's soul as being human. We are of a higher order, a conscious personality, when we can do this naturally and easily, though for most of us it is work as a conscious, willed act. When we could reshape the world in the many acts of giving, of lending the world a new beauty, of creating where there had been a nothingness, we changed the world in the image of our soul. When we created our stories, our music, our dance, our poetry not ony in words but also in our deeds, our cities, our architecture, our care for our environments; then we had learned to bring our human consciousness into this reality. We have been penetrating it slowly, laboriously with many false starts and severe reversals. But that is another mark of our human consciousness. We have been persistent in it. We have been energizing our world in our new image.

How miraculous has been this development! It led us to seek the soul in ways the animal could not. We developed culture, religion. We built retreats for the soul in our monestaries and convents, and worshipped deities of kindness and love rather than the more primitive deities of anger and terror. We became elevated and spread these ideas and deeds to other parts of the world. There was no accident in this; it was the power and energy of our soul. It is in our personality to do this and, when brought to the level of culture, it is in our collective personality as a people. This energy flows from the greatest expressed in the smallest, from an infinity expressed in the self. Through the act of giving, this self is then the energy into the greatest of an advanced world of man. This is a world of individual human being each free to be in the self within his or her mind as a personality. The energy of the soul that flows there is an energy capable of great power.

We mold our world by how we are in it. How we see things, how we touch them or love them or remake them, all are indicative of how we are as a personality. In effect, we lend things our energy when we are in contact with them. When we focus on something, be it a drop of dew or a fabulous, beautiful panorama, it is our energy that flows there. When we gaze at the stars or into the eyes of a young child, we are energizing these with the energy of our soul. The dance of a butterfly or the glow of fireflies are all forms of energy that become trans- fixed by us when we see them with love and enjoyment. We project our energy there. If we meditate on the beauty of a statue or see form in a simple rock, we are lending them how we are in our soul. The soul moves with great power, but we are in the soul how we see these. How we are excited in ourselves by what fixes our attention is how we project ourselves into what it is we see. It is to give ourselves into it and thus to lend it our consciousness of it. We mold our world by how we are conscious in it.

What a precious being man can be. When conscious, to focus more fully on this existence, we can bring out a beauty and greatness of our humanness that has scarcely been achievable before. Though we are in this realm but a short time within each lifetime, how fabulous to be able to focus our attention on this existence and lend it the energy of our soul. We live but for a short and precious moment; we should not squander it on mean and servile pursuits. We are free beings capable of greatness the world has only but recently been awakened to. How futile to wallow in decadence, or to plot destruction and wars. To waste our precious energy on these is absurd, not befitting of man with a soul. Nor is there need to suppress the soul in severe self denials or asceticisms. We are alive. There is work to do. In the short time allotted us in this existence, we should expand our effort in the joy of being, rejoice in our works and cares. Our every moment of existence is filled with the soul. It is an energy that should penetrate into all the things we do.

The personality is the energy that is evident in both the feel- ing of the self and in what is manifest in us as a representation of our greater being. These two facets of personality are what makes us human in a way that the animal is not. We have per- sonality in common but we are conscious of it: "I am". In that self consciousness lies the potential for goodness. We can choose to give, to be kind, to be joyous, in a way that is unique to us. When so chosen, we can then energize these through the energy of personality into our existence in terms of how we are and what we do. When we step beyond ourselves and offer to be generous, to be humble, to be serene, we are adding the soul to our world. This is not done directly from the self but done with the power of identity energized as personality. It is the way infinity energizes for us our world, by projecting our soul into it.

Thus we have the power to project ourselves in all we do. When we love another human being, we are projecting ourselves then. When we love a thing, hold it in our hand, see it with the mind, care for it, we are projecting ourselves with the energy of our being. When we lend comfort where there is pain, courage where is distress, we are offering elements of the soul. And when we overcome arrogance with kindness or anger with understanding, then we are learning to forgive in a way the animal cannot. We are human and as such we do not return anger for anger or pain for pain. We stop it at its source and, if possible, avert it before it even has a chance to come into being. In this is the greatness of being conscious of the soul. We are masters of our destiny and we do not pass on the blame to a lesser being. Same as we do not pass on a violence or willful error, we do not pass on blame. Cruelty and injustice have difficulty thriving in an environment of human beings, for they are not projected by their personalities. When we refuse to pass on hurt, there is great power in this. It is the power of a conscious self image as a soul. To fail to forgive is to fail the power that is our greater human being.

Throughout civilization the personality has struggled to bring itself into our world. We had glimpses of it through our great religions, teachings of kindness and humility, of love. We had glimpses of it when we tried to be just in our laws and judgments, when we abolished human slavery and inequalities. The efforts of these seemingly meek acts have had immense results. But our lessons are not always well learned. There is rampant crime and terrorism in today's world. There is cruelty and poverty and oppression of the soul in many societies. But where we succeeded in shedding these, we had created the foundations for future growth as a civilization. We had learned to rise above old superstitions and seek to understand the universe on its own terms. They are all works of personality universal to the conscious mind. In our institutions of learning, of science and medicine, of commerce and exchange; all express a human soul that has found it more beneficial to agree rather than to coerce and to seek rather than to destroy. We give of ourselves in these pursuits, it is the work of our being here, and from them grows the collective level of our human consciousness. It is not that these achievements cannot be undone; a lapse in vigilance can quickly revert back to a primitive state. Rather it is a lesson that the human soul can be victorious over its predatory past and project itself into a world of care and love and understanding. To love and to forgive, to give, are powerful human traits. To trust, to be sincere, to be gentle and serene are powers that allow a universe to focus in on itself through the personality. Free from coercion, we are then able to pass these powers on to others. When we give, we are conscious. In this is the energy of a greater world.

Chapter Twenty-Four
Within Our Reach

WITHIN OUR REACH lies a greater world. When we seek the soul, reach for the miracle that is personality, the reality that is being, we reach for a world that is beyond mere freedom; it is more than merely the right to occupy one's own space in time, to be in the mind; we begin to reach for the right to be ourselves as we are in the soul. It is to have the soul. A greater world lies within our reach when we choose to seek this soul in who we are, a world of the supreme individual. It is a world where the value of the individual is validated by his or her having a soul. It is a greater order than that man-made. This is the world of Habeas Mentem II.

There is a vortex which of necessity brings us to this threshold. We are obviously more than merely "I am." Though we are free to choose only in relationship to how others are free to choose, through agreement, we are also not free to trespass or coerce, to destroy another's being. For then it is to be without the mind, to be devoid of the universal order. Thus, we are constrained to seek agreement where such exists and thus to seek our individual space in time in the order of our universe, to be within one's identity. We must be in, to have, the mind. We are free to seek oneself, to know oneself, to be as we are in our greater being. Then in this being, when free, we are who we are in terms of how we are in this reality. It is existence, mind, order, and identity all equalled as one. This is made possible by the mechanism of an infinite interrelationship.

Now we are faced with a greater reality: the reality of the soul. This reality is manifest in not only how we are, but also in how we believe. The energy created by this belief, it being related to who we are in this existence as well as who we had been in past incarnations, is the energy that creates for us the settings and circumstances of our current, everyday being. We are in the mind as we believe; thus we are in reality as we believe. Through the energy of belief, through the power of an infinite interrelationship, the All that Is, we create our own reality, how we are, in terms of how we believe. This is the main theme of the second book of Habeas Mentem: "To Have The Soul." What follows is a social reality coordinated by, and whose social institutions of necessity must reflect, the nature of this belief.

Even if the precepts of Book One of Habeas Mentem were socially totally negated, that no human being had the right to seek to be in the mind, the second precept of Book Two would still hold. Albeit, it would hold poorly, miserably, painfully, but man would still be, both in the self and in the greater society, as he believes. Of course, if we were free to seek ourselves and to occupy our own identity, if we were free, then it would be easier to seek out and to live according to our belief. But, though this is the desirable and ultimately inevitable state of affairs, it is not mandatory. Should we fall into an unconscious state of society, belief can still carry us through until such time that the oppression is lifted. The soul survives. If it believes, no matter how arduous reality, its energy will form its individual reality. This is what characterizes man from the animal. We can believe, be conscious, be conscious of this belief and persevere with our will. We can have the will to have the soul.

A free world is infinitely more desirable than an oppressive one, for it allows the universe's order into our lives. But even in a servile world there exists the conditions of survival of the soul. It shows up when a person refuses to partake in this servility, refuses to harm another, refuses to steal or to lie or to cheat one's fellow man; most importantly, he or she refuses to pass on injustice and pain. These are the heroic marks of a conscious being, which in an oppressive world will be sought out and made to suffer and sacrifice. A conscious being can never have the satisfaction of the release afforded to the person who succumbs to that servility, for that is the path of a lesser man. There is no need for martyrdom, but a judicious understanding of where lies the cause of oppression with a quiet resistance to it can lift much of the burden. A being capable of greater reality chooses only conditions that will help others rather than damage them. At times these may be heroic deeds, but there thrives the soul.

When a society is free and the ideas of Habeas Mentem are free to be observed, then there is a natural tolerance towards each other's existence. To seek one's mind, to be oneself, and to seek one's beliefs in such a world is an achievement unencumbered by a systematic destruction of one's being. But in a servile society, the personality that seeks to be one with his or her identity must turn more severely to the soul. If the color of life is dulled, the personality forced to cower and bow, if individuals are made to spy on one another; then the resulting fear and mistrust of necessity forces us from the mind. We become disassociated from our being and our souls suffer. Our being suffers. And our society reflects a drab existence. But there the soul, like a plant that has scant chance of survival on a barren rock in a deep cave, nevertheless finds root and becomes alive. It has the ability to choose, to seek oneself no matter how handicapped it is in this seeking. It is a form of resistance, not social and political, not violent, but of the most positive sort: To refuse to pass on hurt. It is a meek resistance but in the manner of the universe, it is immensely powerful. It is the power of being still. Yet, when we choose the soul, "Who am I?", when we are in our identity, we can move the world.

The mind creates deep within itself. In its strange and mysterious, irrational self, there is the power to become one with our identity in the universe, to merge with the power of the stars. And thus to move reality without apparent, outward motion becomes real. This is the power that gives us life, that allows us to think, that forms for us our limbs, that gives us hope and love and dreams. We are scarcely conscious of this; mostly we are unconscious of it. But it is the power of the mimd that dominates existence. We are, we are our being, our being is in its identity. These are the conditions that are brought about when we believe and thus seek the soul. To have the soul is then to become centrally positioned in this matrix of existence, to become one with the vortex of being. It is the power of a reality powered by the energy of our personality to manifest our greater being in this existence. What form of oppression, of unconsciousness, can long stand against the force of such belief, such faith?

There is a greater world that answers to the calls of a greater force of mind. It is the world that answers to our belief, to our power to do goodness in the face of pain, to heal in the face of evil and fear. It is a strength that appears to come from nowhere but within, yet it moves the world without. It is almost as if there are spirits and gods who can guide us and help us in times of need. Perhaps in a simpler time we could have believed in these. In a universe that has the power to create living beings, why not beings spiritual? Yet, it would be questionable whether they would be better suited to direct events in this existence than through our being. Even nature makes mistakes; why not souls? It is in the nature of a growing universe that there must exist risk and potentiality for failure. The burden falls on us. There is a greater power that holds reality together for us, that shapes it and moves it to meet our sometimes desperate needs. It is the reality of our own making, perhaps looked on by the gods, but an extension of our individual being nevertheless. Through belief, faith, will, and an intense concentration of love, we create our reality from deep within the soul. There, meekness and serenity immensely overpower violence and conquest. There an oppressive world is forced to retreat before a new consciousness.

We are, we are human, we are the soul; these are the mechanisms that are manifest by a universe defining itself through interrelationship. In a conscious mind, we have the soul; we believe. We are a personality. We are our existence. We are the person we love and produce to the world. These are all values we affect in how we believe and how we choose to be conscious as human beings. Whether our personal reality is terribly poor or affluent, peaceful or violent, we are bound within the mechanism of a universe that allows us to be at the center of our identity, to move reality with our soul. If we do not coerce others, and if we are not guilty of being selfishly self-centered at the exclusion of all others, then this soul flows naturally through our being and we materialize it in our everyday existence. We can then change what is wrong in us and work to bring about our desired reality. We work with ourselves through the greater self, and thus have assistance in real terms in how we will achieve what is desired. We are then conduits of the power of the soul and as such energize all we touch with our existence. The terrible poverty of society is lifted; the oppression of a servile world is dissolved. But first we must believe in the value of the individual as a soul.

There is a simplicity to a universe that expresses the most complex in terms of the simplest. This is the miracle of the mystery that is our universal order. It is a miracle that had been praised in countless parables, raised to holiness by religions, interpreted to man by countless generations of priests. The miracle is that when we believe in the soul, when we believe in an order higher than the self, we are seeking the power of our greater being. When we believe, we are as one not only with a simple faith but become as one with a greater, metaphysical reason. When we have the soul, we bring the mind of the universe into our individual existence and thus we glow with the power of a being made real. We become in the mind, have the soul, and radiate an energy of life that illuminates reality with the heightened energy of our personality. When we choose the soul, we become future man.

When we choose to build rather than to destroy, to help rather than to hurt, to give rather than to take away; we are choosing to have the soul. When we seek to forgive where there lacks understanding, to praise where there is failure, and to give hope where there is despair; then we are transcending our everyday common existence and step forth into our greater being. We begin to work with our being, moving reality in a mysterious way from the far reaches of infinity rather than from the immediate reaches of our physical world. It is the infinite strength of silence, the bountifulness of serenity, and the immense energy of standing still. What power is this miracle that moves reality from one's gentleness of being? It is the power of choosing the soul. It is irrational, mysterious, metaphysical; yet, it is individual, indestructible, and the greatness of an infinite simplicity. It is he power of being in All that Is.

This is the indestructibility of the soul. The order of the universe transcends any man-made order, though this order

becomes one with the man-made order when we live through
agreement and goodwill. To love one another is a far greater
force than to challenge one another in competition. To encourage, to help, to be patient, to await result with courage;
these are conditions that could make our man-made social order
even more empowered by the order of the universe. The infinite interrelationships that are reality are also the reality of our
social order. If they appear man-made, man's mind lending
them a rational being, then we must also think and believe in
a man-made order irrational, yet being the basis of human life.
It is that other order, the order of a living universe focused in
on itself in an individual existence, that is called upon when
we seek out the soul. It is the beautiful energy of a mysterious
existence called upon to guide us in our daily existence, the
complexity of an infinity brought down to its simplest: We are,
individually, an image of our greater being. What man made
order can be greater than this?

This is the greater world that lies within our reach. To reach
it, we need but to choose consciously, to be conscious of how
we choose. So chosen, we must choose in a way that gives us
the soul and thus positions us within the reality that is our identity. When so chosen, we have positioned ourselves within the
space-time interrelationships that define our identity at infinity: "Who am I?" . . . "I am the infinite definition that is my soul."
From this definition flows culture, civilization, religion, and the
beliefs that propel for us our individual reality. We are born
into this world on the edge of these beliefs, activate them at
the level of everyday existence with our consciousnesss, and
then release them at death into an existence of their own. This
is how we reflect our soul. How we choose the soul is how
we reflect our identity in this existence. In a roundabout way,
we have completed the full circle of Habeas Mentem: We can
transcend the power of the mind, of reason, and enter the
power of the mysterious and irrational, through the energy
of the soul. We are; we believe; and in how we choose to believe
is who we are. Thus, we are in the mind how we choose to
believe. This is the "who" of our identity. Our reality materializes there.

When we transcend the space-time interrelationship of how

we are with who we are, our will becomes the power of our identity. When we are positioned within our identity in terms of who we are as a personality, we have a universe working with us. It would be, perhaps, natural for us to expect this force to be perfect; it being the force of an infinity. But same as nature has room in itself for error and flaw, so must we assume that the force of infinity likewise has room for flaw. It is not a perfect universe, for it needs to grow. The soul is not all prescient, for there is risk in existence. But there is a natural force that aids us in our existence and guides us with the order and wisdom of a living reality. We then materialize our personality in our everyday existence. Then, in how we will our existence, though it may be flawed and full of error, is how we express our identity here.

This happens naturally, all the time and, when free from coercion, it is free from external error. Our soul is free to be itself, even if imperfectly so. Its "who" becomes evident in all we do: In how we live, how we work, how we treat each other, how we raise our young, how we build; how we pray. All of these are how we pressure existence with how we will. They are the manifestations of the reality we live in. Our existence in it, our work, our world, our art, are all signatures of how we are in the soul. When we become conscious of this, we materialize for ourselves a higher order. We become a greater being in a greater world. This is done simply, individually, humbly, and it is all done in relation to how we chose to believe in the soul.

So we are brought once again back to the individual. Once again the universe's immense complexity is placed squarely on the frail shoulders of an individual being. We are, through our being, the miracle of a universe focused in upon itself. Through us is the mystery of the energy of life made flesh in this reality. We occupy it, we work it, we lend it the energy of our soul. How great is the world in our new, greater existence! And our social reality also starts there. The new individual, manifesting his or her greater being, seeks the soul. The hardness of our cold world is softened by the gentleness of the new personality. Our fellow creatures become less afraid of us as we accept reality on its own terms but grace it with the wisdom of our new being. We do not trespass, do not destroy, do not hurt or damage. The natural gentleness of having a soul is transplanted to the rest of existence, to both the plant and animal kingdoms, thus making our world human in ways we still can but dimly perceive. We are the new man, new woman. Our new society is made of institutions that serve this new being. It is a more joyous world, more playful and humorous, more funloving; but it is also a more just, less competitive and more caring world. It is the reflection of the joy of a universe that can see itself in the material existence of our new being.

In our new, greater world, our social institutions become more a fellowship of souls than a hierarchy of power. The power is one of agreement, of consensus, of exchange, rather than of oppression and servility. It is a new world, courageous in that it trusts itself to a belief in a universe that is great in the body of an individual human being. It is elevated enough to believe in the greatness of a human soul. In this fellowship is a world that becomes free of violence, of mistrust, of suspicion, of fear and theft. To trust our fellow human being, our fellow soul, is a power that our reality has scarcely experienced. Yet, it is the power of a universe guiding its own souls. When two hands clasp in greeting, when we look into another's eyes, when we trust another; we are looking into a fellow soul that has chosen to travel this world with us. How great it is to say: "I worship the soul in you!" It is a force that ties identities from opposite ends of a universe into one. To be tolerant, to be fragile, to be strong, to be willful and yet to be affectionate. It is to be human. When we are conscious of this, when we wish each other well, we are invoking the subtle and yet great power of the soul. It is the power of an identity, a being, an energy of belief, a personality, a fragment of All that Is. It is who we are. In the definitions of a natural order that has defined itself as our physical universe is a value of reality that has defined itself as a soul. When we believe in this, when we believe in the power of the soul in each individual, we can then believe in the natural order that has given us its greater identity. We are made in a greater image. There is a greater order in the world of man because we have a soul.

Chapter Twenty-Five

The Given Word

THE GIVEN WORD UNROLLS in patterns of our creation. How we give our word reflects upon who we are as
individuals. It signifies how we materialize our being, how we
create from within our belief, how we project patterns of our
creation into greater and greater dimensions of interrelationship until we can span all reality with our personality. It is the
mystery of the soul that how we give our word, how we speak,
how we agree — are all how we materialize the reality that
reflects the personality we are. We are the authors in ways we
cannot yet understand; yet, how we name things, how we touch
them, how we see our world and admire it or communicate
with it through our being, are all how we create from within
ourselves definitions that set forth a separate universe. It is the
universe of our new being, our consciousness, our new identity. How we give our word is, through the infinite manipula-
tions of reality, how we manifest our being. It is who we are.

We had long given words to things, named them. From the
dawn of our existence as man with a language we had given
words to help us understand a mysterious world. The names
were probably a form of magic, meant to evoke forces of a
greater reality. But we had seen that now we need not name
things in the way we had before. The universe already names
them for us, through the definitions things have from their infinite images of interrelationship. Each thing is how and where
it is because of how the image of everything else, All that Is,
has allowed it to be. Things already have "names" in a universal way. But through belief, we can now also give names to our
mysterious reality. Through the spoken word, the word given

193

in the way of truth, we define reality with our new power of being. We are, we are conscious, and in this consciousness, we create our own reality with our soul. Patterns of reality unroll from each given word.

We create our reality. We personify our existence. But to do this we must be as one with our existence. When we do through agreement, are gentle in all our dealings and do not use force, then we are positioned within our identity, in balance with our soul. When we are sensitive in our being, when we cherish the beauty of the human soul, love our fellow beings, we are calling upon the power of belief to create our world in our greater image. We are calling upon the courage to do goodness rather than to strike back with fear, to give rather than to greedily take, to energize and construct rather than to tear apart. We are calling upon Creation to render us human rather than merely humanoid with an animal soul. To rise above the instinct of anger, to focus one's existence within a consciousness of belief, to believe in a greater being, to tell the truth; all are elements of the being that defines for us who we are as an identity. We are in this reality as we believe, as we give our word, as we have the courage to be gentle and kind. These are immense demands on the man of today, but it is the condition that transcends us from our primitive past to the future. To be free, we must justify the sanctity of our individual existence. We can do this only if we have the soul. To have the soul we must be within our identity, our greater existence, and be conscious of this. To do this we must seek out agreement rather than force. When agreed upon, we must obey the given word. To believe in this is to create a new reality in our world.

These are the conditions of future man. It will be upon these conditions that we will be judged as a planet, as a collective of individual and free human beings with the soul, who had populated a difficult and yet beautiful planet and made it a conscious world. How great are these goals, and yet how distant they seem to us from where we stand today. With confidence and care, with determination and singleness of purpose, these goals can be achieved. We have named things for eons of years, and now it is demanded of us by reality to name things not only with our words but with our soul. To tell the truth, to speak consciously and with reverence for the greatness of our world, is to give it a new meaning in the universe. It is to give our planet, Earth, its own identity.

To have the soul, to be who we are, is an extremely important manifestation in our reality. It is how is expressed our personal identity in this reality. Through eons of generations of existence we have come to the present. Through deeds and words and thoughts that stretch back to our soul's beginnings, we have modified reality with the presence of our being. We have tamed worlds,or made them wild. We have healed with a gentle being, or spread pain and destruction with a ferocious one. It is all written on us in the patterns of existence we leave. We are those patterns, exist in them daily, as the patterns of our soul penetrate into all the intricacies of all we do. Our individual being is a signature of the personality we are in our greater existence. When we become conscious of this, the energy that is our being grows.

The spirit of man grows and is nourished, or it is starved and it perishes. How we are, how we do to our fellow beings, how we do to the animal and plant forms who are our lesser beings, how we do to the physical reality, all reflect how we are energizing our soul. When we are conscious of the self, we grow. When we are confident of our being, are not afraid to trust to fate and to do as we believe, our soul grows. When we are cautious and careful of all we do, of how we promote goodwill and a sense of well being, our soul grows with us. And when we are truthful, when we have the courage to trust one another, when we are powerful in the mind in ways we but dimly understand, then our soul is growing to its fuller potential. We can be great individually as souls if we have the courage to seek this greatness, to believe in it, and to trust in a universe that can move immense power by being still. To seek beauty is to nourish the soul. To do kindness is to bring forth the soul. To have the soul, to be as one with one's reality, is to exist in a special existence that works with us in ways that we can still call but a mystery. In time, in our conscious minds, it will be revealed. That is what is meant by growth.

How could we have been otherwise? We did not know of our greater existence, did not believe in the soul. When all things were explained confidently by an appeal to science, to be truthful was to be handicapped, for one had to be caught to be proven false; it made no difference how we were in our existence. Without the soul, we were as barren as the rocks of the desert. But even there life struggles on and becomes fruitful under cultivation. We are human beings with great potential in the universe. We have souls that, when fully developed can move worlds. We are creators of reality, when we become aware of this. To have been so small when we are so great is to be absurd. Yet, how were we to know to be otherwise? We did not know that to love one another had a force far greater than we had ever imagined. But we must start in small ways, to agree, to help, to do in ways that do not harm, to tell the truth. All of these create patterns that are beneficial for the future development of our soul.

The greatest is defined by the finest; the strongest is defined by the meekest. The gentle energy of man is the power that propels the soul into being in our existence. When the mind of man perceives this and can believe it in the way it can believe in its own existence, "I am", then we become as one with the mind, become in the mind, and the power of the universe begins to work with us. It is a concentrated effort, not easily kept on course, for we tend to get confused and lost in our random motions. But it comes naturally when we are in the mind, and we are in the soul when we become conscious of this and begin to seek it. To believe, to seek a greater being, is an important element towards achieving the energy of the soul in our everyday existence. Because we create our reality in our greater image, the image we project into the world is the existence that will be reflected to us. To seek to avoid violence, we need to avoid it in ourselves and then seek a reality that is not violent. To seek to avoid disease, we must live a life that is healthful and seek health in our souls. To seek out loved ones, we must be able to love others and seek out this love in others. We need not be afraid of achievement, for to seek achievement is to believe in the power of the mind when it is in the soul. It is to have patience when there is haste, to have courage when there is fear. It is to offer solace when there is sorrow and to offer encouragement when all seems to be despair. It is to have the soul to rise above the confusion of the present and to become conscious of being in the mind. To seek reality with the soul is to become confident of one's existence by appealing to the meekest forces which yield the greatest focus of existence. We are in our everyday being as we are in how we believe in the soul.

Reality moves in and around us like a sea of being that fills every crevice of our existence. This is the energy that is the being of our soul. It is the light that shines from our eyes as personality, it is the handsomeness of our limbs as well as the melody of our speech. It is the creation of the artist or the power of the actor. The soul is intertwined completely with the forces of our being and of personality, forces into which we are born and from which we die. We exist on Earth but for a brief moment, yet how rich and complex our lives can be in that brief span of time. We can have the luxury of loving our children, of teaching others, of building great works or of creating with beauty and simplicity, of being humble. Our soul speaks to us from the millions of things we do that betters existence rather than damage it, that elevates and energizes rather than dulls and tears apart. It is a brave world that can toy with the forces of darkness and destruction, but how much more beautiful and courageous if it can seek out with the power of light. To be conscious of ourselves, of each other, of what and how we do, is to become conscious of who we are. "I am Man" is a powerful force to release into our reality when we release it with an energy of being in our greater being. "I am the Soul of Man" is an even greater force to evoke for it is to seek with truth and love and ultimate goodwill to the fellow souls who are traveling this planet with us. It is to evoke an energy of the given word as it is spoken by a new, conscious being.

When we seek with friendship, when we love our fellow man, when we love our family, our wife or husband, our children; when they love us; we are sharing our soul and in our greater image. We are focussing on the greatest force of the universe and are in turn the focus of that force. We become positioned within it, it works with us, and we become filled with the energy of its being as it flows from the vast reaches of Infinity into our daily lives. To Love is to become as one with All that Is. It is to alter our identity of who we are in the image of our soul as we are within All that Is. To love is to evoke a great force of our universal reality; it is to bring forth into reality the fruits of a Supreme Greater Being. It is a positive force of kindness and caring, of goodwill. It is an immense force that changes a world. When we become conscious of loving one another, we energize our world in ways that to us is still but a distant mystery. It is in the world of the irrational, of being greater than can be understood by the mind, of a power that is latently possible in each one of us. When we are able to love, we are able to energize our reality with the energy of a greater being. It is so simple, though it is infinitely complex. To love is how we gain the soul.

The power of light versus darkness is brought forth into our world by the Given Word. To Love is to bring the power of the Soul into our world. It is to nourish life, to enrich our existence, to give health, to have faith in our given reality. When we seek with honesty and friendship, when we give our word and keep an agreement, we are empowering our reality with a force of existence that is moved by All that Is. We invite this greater force into our being, into our lives, into our society, into our nations and kingdoms and social institutions. How much work there is to be done by so simple a force. Yet, how powerful it can be. It is the new name we can give our world, the "I am Man" planet that awakens within the universe. The energy that flows there is the energy that gives us life and is the creator of souls. To be, to be human, to be gentle and kind, to be conscious, to love; these are the values of the new identity of man who has the soul. To have joy, to laugh, to dance, to sing, to have the courage to be free; these are the elements of a new soul who has embraced life on its own terms and thus enriched his or her existence. To be creative, to be serious, to be confident, to be lighthearted; all are characteristics of a being that is molding the world in one's greater image. To be pleasant, to be generous, to be courteous; they are the marks of a greater being. We invite the universe to do with us as we do in our minds, as we are in the soul. It is to see the world through new eyes and to love it dearly. It is a beautiful planet when it is full of the magic of All that Is. It is a world of Being: to be Man.

In each lifetime we will have the opportunity to exercise our being in terms of who we are. We will have the chance to seek out friends, to seek out loved ones, to build relationships that may last eons, to do great works, or to care for those who cannot care for themselves. To have the soul adds an immense dimension to this opportunity, for it offers us the courage to choose carefully how we will be in this soul and thus in this existence. How we will manifest our being in this lifetime versus how we had been before; how we delight in the discoveries of each new existence or how we delight in and cherish our friends; how we hope and dream to be; all will leave upon us the mark of our signature of who we are as beings. In future lives, this will be our new reality. How great is a universe that can, from the first interrelationship of three, become so great, so full of imagination, as to create existence with a soul. The soul is a separate force of existence. When we have the soul, we are the new man.

To Give, to Love, to seek with Truth, to Care, to make Beautiful, to bring Joy; these are images of the name we give to Man. It is the images of the Soul, the patterns that unroll from the Given Word. To have the Soul: This is Who we Are.

Chapter Twenty-Six
Let Us Work Together

LET US WORK TOGETHER to have the soul. Let us gain the soul in terms of who we are. Let us be conscious of our personality, of who we are as physical beings, of who we are as personified energy. Who are we as an identity? Who are we in the infinite interrelationships that energize and materialize for us our being? Who are we in the mind? When souls meet, these questions are asked of necessity of one another. When our personality-reality images merge together in friendship, we work together to bring together our respective realities and thus bring our souls into our reality.

We have seen how it is that we have the soul. When we have kindness and care and charity, when we are humble, when we forgive, when we do not trespass on the being of others; then we have the soul. It is an energy that flows from our being, from our greater being, from our distant pasts, into our present mind and consciousness. Through millenia, through myriad lifetimes, the soul has gained consciousness in this dimension of existence. We can look back upon our universe and wonder: Do I have the soul? Can the soul be part of my identity? Is the soul a conscious being in my reality? We move and energize reality in ways that can never be understood by the rational, conscious mind. It falls into the domain of the irrational, of belief, of prayer, of meditation and of a reality from that other side of the mind. It is the reality as seen through friendship and love.

We have a great power when we have the soul. It is the power of moving reality with the energy of our being while being still. It is a discipline that harnesses the power of anger or rage into a force of love, of agreement, and of mind. We have a great power when we have friendship, for it is the pure energy of the soul. An individual being is the focus of but one, but in friendship there is the power of two. With two realities supporting one another we invoke the power of reaching out into infinity, as in prayer, to move reality from our greater being. When we reach out with love, we are engaging an even greater power of the greater reality of personality, the "whoness" of our reality, combined with the personality reality of another. These are dimensions of infinity, of powers and life giving energies that are far beyond the comprehension of our mortal minds but which, through the extrapolations of an infinite interrelationship, exist within the framework of an infinite reality. Our souls are the products of these, going back to the beginnings of time, formed in the early crucible of a new universe. We are real figures in the universe, as real as the identity that is our being. Our souls are real entitites. When two people meet, these real souls are brought together.

Because reality is moved by how we are conscious in our greater identity, how we choose, how we move reality with our being, how we do, how we are in relation to our fellow beings; all are important to how we are in the soul. To become a greater person, to rise above one's personal chaos as well as the chaos of a still semi-conscious world, is an achievement of immense work and will. The soul is not gained automatically for it requires a conscious effort. The universe helps us arrive at consciousness, but it is our identity's consciousness that helps us arrive at a "consciousness of consciousness" to gain the soul. We must know how to choose with friendship and love. Our consciousness is evident in all we do.

It is easier to gain the soul when we work together with love. Though we are moved by our reality, forced to respond to all of is circumstances, we are also the authors of these circumstances. The collective energy that through the myriads of incarnations defining our being has formed itself into the definition that is now evident in our flesh and blood, our body, our personality. From distant and obscure points of consciousness we have been gradually refined into our present being, our person with our personality, and our respective reality. The threshold of consciousness had been crossed only recently when we became man. Now the initiative to carry it further past that threshold is given over to us. We must want and be able to choose. We must seek agreement with our mind's reality. We must seek agreement with the reality of others. And we must learn how to position ourselves within our reality to gain friendship and peace and compassion in ways that can position us within our soul. To give, to forgive, to be caring in all we do is to love one's being and that of one's fellow man. It is to have love for our reality, for our fellow creatures, for the planet's reality we inhabit. This is how the personality energy of our soul is manifest in this world. To love one another is to help manifest this soul in this world.

The spiritual reality that is expressed in the consciousness of love becomes the new reality that moves us and that is moved by us. We are the authors of our reality. We create reality in terms of how we believe. We believe in a greater reality of the soul when we act through love. This new spiritual reality is then what moves for us our world. It is the world that responds to prayer, that invokes the energy of souls, that forms for us the bridges of kindness and understanding that help perpetuate this new world of the soul. It is a world that had been long in forming itself through the countless generations of our being. But we are conscious now; we are aware of what it is we do to ourselves and to others. We are conscious of the majesty of loving one another. The energy of this new spiritual reality is like a beacon that guides us into the future and opens new dimensions of reality. Yet, we cannot enter these new dimensions until we have the soul.

When we are no longer in conflict with our existence, when we are no longer hostile and can avert hostility with the energy of our being, when we can radiate a confidence of existence, of being in the mind and of being masters in our own realities; when we no longer trespass on the realities of others, when we no longer deceive or sow dissention, when we are no longer fallen and have lifted ourselves into the world of a new consciousness, when we have the mind: We become Man. When we can help our fellow creatures, when we can help reality by shaping it in ways that are more beneficial and ease its burdens,

when we beautify; when we touch our world with an inner compassion, when we love the truth, when we seek to make life
a litte bit more beautiful; when we rejoice in another's happiness and when we help to make another happier, when we
do with love and love our fellow man: We gain the Soul. This
is when we create a new reality, a reality that can work with
us and one that is accessible to us when we are in the mind.
When we do this, we are both the creations and the creators
of a greater consciousness that is found in the soul. When we
do this together, work together to bring a greater spiritual reality
into our world, when we no longer trespass on each other's
realities but rather find companionship and friendship there,
we are bringing together our respective infinities and our souls
meet. A great void becomes filled with the joy of our individual
being and our creation, the brilliant energies of our personalities, and worlds touch through our physical beings. When
we meet in friendship and work together and love one another,
universes meet.

We live on a beautiful planet filled with great energy. It shows
up in the formations of natural wonders, it becomes evident in
man's many faceted cultures, in the great societies we had built
up over the millenia. We capture this energy in our works of
art, in our architecture, in our lore and traditions, in our forms
of worship and beliefs, in how we are as people. If we are
hospitable , trusting, faithful and truthful, then this energy is
abundant in our society. The tools with which to move reality
become known to us, our sciences prosper, our knowledge of
the universe expands. The planet's energy becomes evident in
the works of the hand, of what is written, of what is formed.
Hands of friendship extend across the face of the planet and
the energy of the soul becomes manifest. It becomes a brave
new world, one with the courage to believe in one's fellow man.
It is the courage to believe in the power of the energy of the soul.

It is a brave world that can believe in the soul. It requires
the courage to trust another person, to have faith in one's personal reality. It takes will to negate the actions of a liar, to identify
the thief. It is work to stop a trespass before it is able to spread
like a cancer, infecting all it touches. There is no reward in forcing disagreement in the word of the soul, nothing is gained

through conflict or in sowing dissent for these are negated by
the soul. But it demands a conscious mind to bring this into
reality. If we cower before aggression or fail to have the wisdom
to prevent it altogether, then we are unawares. If we are im-
pressed by another's illicit gain or are readily flattered by deceit,
then we are failing in the soul. We cannot be weak, for to believe
in weakness is to not chance the risk of defeat, to be slavish.
That risk is mandatory for the conscious mind, for we cannot
be conscious and free without the courage to fail. We must
have the presence of mind to force into failure what will steal
from us the soul, what will seek to trespass against the mind,
what will force us from agreement. To be conscious is to allow
only that which elevates the soul that helps agreement, that
nourishes friendship and love. To be conscious is to refuse to
pass on pain or deceit or coercion, it is to reject oppression
of the weak, to disarm injustices that weaken the soul. It is work,
it is to help rather than to hurt, it is to lift up where others had
fallen. We can do it together.

But it is a braver world that can seek to exist without the soul.
To establish a world whose greatest legitimacy to order is the
rational intellect of man is to attempt to transcend the power
of friendship and love. To be without its greater, spiritual being, to be without the mind, but rather to be entirely within
a manmade order is to negate a greater, universal reality. Then
the mind is but the product of an accidental probability that
had become man and saved from chaos by man's mind. There
is no room for a greater being and thus there is no need to
answer to any greater order than that formed by man's social
order. The ultimate achievement of such a mind would be a
totally scientific, social order to which all would answer and
from which none could stray. It would be a total worship of
the rational mind embodied in a totally benevolent social order
based on the linear logic of a scientifically designed world. But
this one dimensional vision of reality fails to perceive an infinite
interrelationship. When we believe that there is no greater order
than one manmade, we close in our universe within that man-
made creation. To believe only in the rational mind of man is
to negate the irrational, to ignore interrelationship, and thus
to establish a worship of the human brain. The universe

becomes but a vast random, accidental order that has no legitimacy in a human social order. And the mind of man is forced into the patterns established by the rational side of the brain. It fails. We are more. We need the freedom to expand and stretch our being into the ever expanding universe that is our soul. The rational mind of man is but an introduction to consciousness, not an end in itself. To believe in the rational side of the brain at the expense of the irrational is to put our future development entirely in a faith of the means of this development as an end in itself. There is no harm in gravitating about this point for a time, but it will eventually become oppressive to us. The rational brain is not consciousness itself, only an opening to it. There is a greater universe that has an order that is far greater than any order manmade. We must have the courage and the freedom to lift off from that comfortable level of rational achievements and explore what is still mystery to us. The dictates to the mind, when we are in the mind, are from the infinite. When we act through agreement, when we care and love one another, when we are generous and have the soul, we are able to lift ourselves from the heavy gravity of a linear universe and reach into the greater dimensions of our mind. We can rise into the reality of an infinite interrelationship and seek our identity there. To believe in the soul is the easier course, for the legitimacy of the mind then rests with a greater order. Anything less becomes but a tyranny of the mind.

We energize reality with our being. We are the points of infinity through which intersect the patterns of our greater being that occupy this space in time. We are. We are the personalities that lend the human value to this small planet in this corner of the universal reality. To love one another is how souls come into being. If love is the power of cosmic worlds coming together, why would we deny this joining of worlds by denying the existence of spiritual beings? Why not believe in the soul? Why not have souls work with us? It is how we energize reality. It is how souls meet.

There is much work to be done to lift us from our still coarse existence into the domain of the soul. As we awaken one by one, each one of us in our own way and at our own time, free-

ly, by our will, we will radiate our personal energy, our personality, into our world. Think of the miracle! The great works
of man, of the arts, the sciences, literature, music, dance, poetry,
charity; think of great teachings, of technological advances, of
commercial enterprises prospering in an environment of agreement and exchange; think of advances in medicines, in communications and travel, or the explorations of distant worlds
and of meeting beings in other parts of the universe; think of
theatre, of architecture, of history. How great are the works of
man's soul! To work together. To care and to trust one another
would allow great advances. To grow in stature over the small
and evil vestiges of our planet's past, to overcome these with
future forces of love and the power of believing in one another's
souls. To have faith. To be truthful and to honor the word given.
What brilliance still awaits us? What can we expect from a planet
that has anew embraced a belief in the soul to create a new
reality? This is the threshold of consciousness to which we are
but now awakening. The oppressions and fears of the past must
be overcome by these. To cross over into this newer world, we
must choose and we must believe. We must trust one another
and care. It is work that we must learn to do together.

As we each pass through our reality, we seek each other's
presence to form our greater, respective realities. We are
sociable and in this is how we define for ourselves this world.
Through how we do and how we do together, through whom
we meet, is how we define our souls in this reality. We had
passed here before and are likely to pass through here again.
In each passage we will create reality in our greater image which,
in turn, will affect us in how we had created. In each passage
we re-activate our reality and pass again through the paths of
our creation. We do it for ourselves; we do it for our children
and for our loved ones; we do it for souls newly met. The definitions we leave behind will define for us how our world will be
for us in the future. The judgements of whether we do it well
rests not with us but with our greater identity and its universal
order. This is a mystery to us, irrational to our rational mind.
This is where souls meet and shape reality. When we meet as
friends, when we love one another, this reality becomes greater
in terms of our being. We become conscious here as souls. This

is where souls meet and are nourished into greater beings. Why not meet in love and peace?

It is the stuff that dreams are made of. It is irrational, but it is real: To believe and to love one another.

Chapter Twenty-Seven

It Is Important to Have the Soul

I T IS IMPORTANT to have the soul. It is how we express our personality. It is how we express our greater identity. It is how we occupy our own space in time. It is how we occupy the reality of the definition of infinity as it is interrelated in our being. It is how we choose. It is how we are. It is how we are conscious of this. Then, we have the soul.

When we manifest our reality with our being, we have the soul. When we pass through this existence with reverence and concern, with love and respect for being, with awe of the divine mystery of existence, when we nourish life and avert pain, we are reaching for the soul. When we succeed in moving our personality and personal reality with the energy of our being, with the force of our conscious presence, when we move reality by being still, we are approaching the soul. When we radiate a spiritual brilliance, when we attract what is joyful and lifegiving and repel what is harmful and hurtful to living beings, when we do this through the natural power of our being, of being in the mind, we have the soul. We then radiate our personality into this existence, we mold this physical being into the image of our greater personality, our greater being and identity, and the mystery of an infinite universe that has made man in its totality image is then manifest in our individual being. We then have the soul, for we had gained it.

In simpler times, this would have been understood in religious terms, cloaked in mystery and magic and spiritual revelations. But in these modern times, there is no longer need for this. It can be seen as but a natural extension of a universe forming itself. The soul is the basic, creative energy of a greater reality as it materializes itself here in terms of a growing infinite universe. It is part of the creating of reality. When we have the soul, that growing and creating of reality is reaching into our world.

Being is the essence of existence. It is. It is how we are materialized here within the vast matrix of an interrelated infinity of energy that is our universe. Being conscious, with a soul, is how we are materialized here as an image of that vast, universal energy. It is the irrational side of existence. Where the three dimensional world of physical being is sensible to our rational mind, the multidimensional interrelationship of the soul is what is an irrational mystery to our mind. But it exists because of our being. We are born and we are. When we choose to believe in this, we are activating the mechanisms of this mystery to bring it into our world. In effect, the soul, though it is reaching for us through the manifestation of our personality, is not evident in this existence until it is brought forth through the energy of belief. Then, through our being it is able to manifest here as a separate force of existence. The world changes then.

Because it takes belief to have the soul, it is important to have the courage to believe. It is easier to be skeptical, to reject ideas until they are proven in a scientifically correct way. To the rational mind, it is harder to accept belief, for it is unscientific and irrational. Yet, this is a necessary paradox. If the idea of metaphysical interrelationship is carried forward to its rational conclusion, when it enters the human domain, rational scrutiny fails it. We are launched into a world of understanding by faith, of understanding in terms of how it is that an infinite interrelationship materializes our reality for us. Because this infinite interrelationship exists but is out of our rational control, it becomes an act of belief. We can only understand it by becoming one with it in occupying our own space in time in terms of our identity in it, to be in the mind. Now we can also be in the soul. If we do not reject the irrational and have the courage to step beyond a scientific need for proof of the soul, then we have the courage to reach out with belief.

Proof is needed by the rational mind and must exist for the sake of our three dimensional reality, whereas belief is need-

ed by the irrational and must exist for the sake of a multidimensional, greater reality, as in infinity. For us to make the transition from seeing reality from our point of view to seeing reality from an interrelated point of view took an act of faith, though
it was explained in terms of the rational, an infinite interrelationship. Now, for us to make the transition from understanding reality totally in terms of a three dimensional universe to
one that is multidimensional, we are again posed with a choice
we must make. Either we believe and have faith in the soul,
and seek to understand in terms of the irrational; or we have
faith only in the mind and seek to understand only with the
rational. Both succeed in bringing the universal reality into our
midst, when we are in the mind. But where one is growing and
creative, powered by the energy of the soul, the other is static.
One is a spiritual reality, the other is a physical reality. Where
one already exists, the other is still in the process of becoming. To become as one with the soul, we need to believe. If
the universe is in fact a spiritual energy, though we cannot
understand this in the rational, then it is suitable to seek to
understand it in terms of an interrelationship of energy rather
than mass, of an interrelationship that is more in the domain
of belief than of understanding. There is no proof of this, but
it is the step that is needed to gain the soul.

To gain the mind means that we are free from coercion and
thus are occupying our own space in time and the universe's
order is working with us. However, though this is more desirable
than a world of coercion, a world where beings are disassociated
from their greater being by existing in a state of controlled
chaos, it is nevertheless not a creative force. It is a state of being defined by being itself. But it is not defined by the energy
that is defining for us life. It is in reaction to rather than the
condition for future growth, without imputing into reality our
being's conscious energy. Though we are conscious as beings
and aware of our existence, we have the mind, we do not yet
have the soul. To have the soul, we must be conscious of the
creative energy of our existence. We have to add to the universe,
to create with our minds what had not existed before. We need
to lend to reality the soul. The mechanisms to accept this new
creative force in terms of infinity already exist. They are the in-

terrelationship mechanisms of the metaphysics of being. This is why we must believe. For when we believe, we are calling forth this new, creative energy of our existence. Then, we are improving not only over the chaos that results from not being in the mind, but we also improve on the order of being in the mind by giving it a soul. It becomes a creative world powered by the energy of love and joy.

Think of a world that is no longer powered by coercion, no longer ruled by fear and force, but rather where human beings are free to be themselves within the Law of Agreement, free to have the mind. In such a world, the legitimacy of human freedom is that we have the mind, that we occupy our mind physically within the space-time interrelationships of our greater being, of our personal identity. But that, in and of itself, is still insufficient. Why should we be allowed to occupy our identity? Why not occupy another's identity? Why not the identity of a superior being or superior authority or supreme master? But if the supreme power is one that is other than man-made, the legitimacy for freedom is carried one step further. We are not free merely because we wish to be free, in effect because we had chosen freedom; rather, we are free because a supreme order, other than man-made, demands of us our freedom. This freedom is demanded not because the universal order that results from it in this reality is superior to the controlled chaos under which we now exist, but because it is the condition that is necessary for us to gain the soul. We cannot have the soul if we are not free to choose to have this soul; we cannot be free to choose if we are not free to be in the mind. Only then are we free to believe. Thus, to have the freedom to believe, we must have the freedom to seek the soul. Without our freedom to seek the soul, our right to freedom is still open to doubt and questionable by a potential, self-styled supreme master. Do we really need to be free? When we have the soul, the answer is a definite "Yes!" This we must believe.

We live and we die. Within this short life cycle we could each be made slaves and never know freedom. If we had no soul, it would not matter, for we would not know the difference. If the soul did not exist, there would be no will, no universal manifestation of a desire for freedom instilled in each, short life span. Within the identity that endures beyond death, there would be no need for human freedom other than as wished by a personal desire. But personal desire is not arguable, since it may or may not be justified; and by whom? A desire for freedom may be but a fancy and wish that should be quelled as soon as it is able to express itself if the hard reality of the greater social order demands it. It is an arguable point, one that leads to a social form of slavery that is superior in terms of social efficiency to a system of freedom. Arguments opposed to the desire for freedom can be so convincing that to argue contrary to them is to entertain the absurd and to be irrational. But that is the key! It is in the irrational that lies freedom. The right to be in the mind is still in the domain of the rational, but the need to be in the soul is already in the irrational. It is there that lies the demand to have the right to be in the soul. And it is also there that it is important to have the soul in order to gain that right. There lies the mystery of our existence. We live and we die, and there is no rational argument for this. But this is understandable in the irrational. We live and we die, and when we are free to gain the soul, we live.

Thus, there lies the key to our legitimacy for freedom. We live and we die, and in this cycle is the root of our existence. A slave is not allowed the soul and must exist for the sake of the master's needs. An individual expression of identity is superceded by the totality needs of the overlords, even if these be a social majority. But a soul is an individual that stands alone, supported only by the belief in one's greater identity in the universe. If the totality is greater than that of human overlords, then there is hope. If this totality is a universal order whose greatest objective is to create life in its own image and to give this life an individual being with a soul, then there is the rationale for this being's freedom. The being, if he or she believes the self to be free, is free. This is chosen through belief and justified only in the irrational. "I am, I live, I die, and yet I am." To live and to die within one's identity is to live and die as a soul. It is to be a conscious being as defined by a universal reality of infinity. If it is an infinity of interrelated energy of love and spiritual being, if we believe in this, then we are free. For then, it is who we are.

To pursue our daily existence with a belief in the soul, in spiritual forces in our daily being, is an incredibly enriching experience. To experience life with the perspective of belief that in how we do things, so is it done with a universal infinity, and as we do with love and care and kindness, so is this added back to our soul. It is to unlock an incredible reservoir of energy of joy and well being in our lives. To believe that we pass through here but temporarily for the purpose of enriching the soul and, in that process, for the purpose of enriching the reality of existence as it is experienced by the being of that soul in the flesh, to make existence so much more meaningful than merely to exist to eat and procreate. It then means that we are working with a universe working with us! Imagine how it is to be tied into all of existence and to become a living part of infinity. To work with creation to add to the energy of a universe; to add this by adding to the energy of our soul. What a wondrous existence! To live a life, to seek love, to find joy in all around us, to find each other as souls, to worship the sacredness of our beings, is to worship the soul in each one of us. To help, to build, to heal, to lift up and improve, to make more beautiful, and then to die with a soul greater than when born. How greatly improved would be our world as we improve our personal identity. As we do, so the universe does with us. As we do as a soul, so does the universe make our identity known in this reality. We personify existence when we are in the soul. We are in the soul when we believe in the soul, when we are free within the mind and when our world becomes free with us. Yet, it is all irrational, for it lies in the domain of belief. Is it not worth the risk? That is where our future lies, if we believe in the soul. It is the ultimate justification in human freedom, for then we can no longer believe in any justification to trespass on another.

Everything is connected, from the state of health of a new born to the hopes of the infirm ready to die. From the love of a planet for its living beings to the love of a parent for his child. From the life essence of a microscopic creature to the life patterns of an unrolling universe. The progression from life to life, death to death, is a progression of souls. It connects all existence into a unified whole. It is in the interrelated webs of this whole that is formed the crucible that defines for us the soul, and it is into this whole that upon death the soul is returned as a new living identity. We are born into the world capable of the soul. Most will live unawares of it and some will die having lost it. But, conscious of our being and of the freedom to choose the soul, to believe, to risk the irrational, we can be the individuals who are born into this reality with the personality that has gained the soul. Then we can die from it with a soul capable of eternal being. To love, to care, to be truthful, to be careful, to be beautiful; these are the elements of being free, of the irrational, of having the soul. All are connected into an inextricable, interrelated whole. It is important to believe in the soul, but this belief is of no value unless it is chosen freely, of one's own free will. That is always the ultimate test. Whether or not we are in the mind is evident upon whether or not we trespass another. Whether or not we are in the soul is evident upon whether or not we are free to choose the soul, to believe. To have the freedom to believe is the final connection of human freedom.

It is important to have the soul, for that is how we create with a universe creating with us. It is important to avoid that which damages the soul and it is important to seek out that which enriches it and beautifies and strengthens it. It is important to being free.

We live and we die. It is important to guard the soul. It is the key to the universe.

Chapter Twenty-Eight

It Is Important to Have Faith

IT IS IMPORTANT to have faith in the soul. It is how we legitimize our demand for freedom. It is how we gain the right to choose to be ourselves, to occupy our own space in time identity, through agreement. It is how we do through agreement. It is how we become our greater being, through faith in the power of the soul when conscious at the center of one's existence, to move the universe. It is to move from that other side of the brain, to create from the irrational what is real and rational, to believe in existence. When we believe, when we have faith in the soul, we gain the soul.

To have the soul is to surrender to the logic of being in the illogical. It is to invite into our everyday existence the power of the infinite being interrelated into a conscious individual able to choose to believe, to have faith. From an infinite identity is pressed into this existence the question: "Who Am I?" The answer lies in our sanctity of being, of love, of belief. In "who am I", when expressed in terms of agreement, is the immense movement of the universe here as it moves in relation to an individual being. It is to move with the power of the soul, of the universal being as that power is moving itself. "I am that I am", is to look into the being of the mind of one man as that power is focusing itself there. To become conscious of it, to believe and to have the courage to believe, faith, is then to bring that power into our everyday existence. This is the miracle of being in the soul. It is a new kingdom where every man is a king within his person when he or she is free from coercion, when they live by agreement. Then, the universe rules there. To have the soul is to become free to love one another.

To reach out with kindness and helpfulness, to have faith in their strength, to reach out with truthfulness and trust, is to reach out into the universe and far back into the miracle of the soul. To believe in trustfulness and in kindness is to believe in the sanctity of our being. How much did the ancients know when they said to "love one another"? The temple of our being is already here, in the individual human being, our own personal being. Do we shine from within with love and kindness and friendship? Are we more than merely animism, unable to give? These are the marks of a person with a soul, he or she who has faith in their soul.

To have faith implies hardship. It is not easy to believe in our turbulent world. We are victimized daily, either by anger or by mistrust or by violence. Our belief is daily tested. We are tested when we suffer. Do we submit and retreat or do we stand fast in our belief? Are we conscious or weak? Is it not easier to hate than to love? Do we have the faith to move our reality with our soul, or do we merely accept that to suffer is part of our turbulent world? There is a vast void between faith and submission. Yet, when we believe and have faith in the soul, we fill that void with our everyday existence powered by the mysterious energy of our greater being. It becomes filled with the energy of a soul moving our personal reality in terms of our greater image. To be in the mind is to have the freedom to have the soul. This is an important first step and one which lends legitimacy to our demand for freedom. But to be in the soul is to believe in the power of the soul in our everyday existence. This lends legitimacy to the universal immortality of our being.

The soul expressed in the flesh, in our being, in our personality, in "who" we are, is the definition from infinity of our being, our identity. It is the definition into which we are born and into which we die. It is in relation to how we choose existence and in how we will choose within it. It is how we love and are loved, how we do and how it is done to us. Yet, ultimately, when we are most severely tested, it is how we believe and how we have the courage to believe, it is with faith in the soul. Through time men had built temples to the soul and to this faith. From ancient times to the present men gather to pray, to hope, to sacrifice and suffer in order to have a better life, a better world.

The ancients offered sacrifice to aid them when their faith faltered; they beseeched their personal gods and lavished them with fetishes to curry favors and privileges from them. But these are false and unnecessary. The power of the soul is in their belief in it, not in some odious sacrifice. The power is in the strength of the human being to be able to resist temptation and to not succumb to irrelevant and cowardly acts. It is to stand firm in one's conviction, even when all around are crying to retreat, to give in. It is to believe, to suffer with this belief and to power with the soul. We are powerful when we are in the soul. When we are without guilt of coercion, when we are truthful before our existence, we have a universe behind us. There is the power of the soul, not some minimal efforts of capturing it by sacrifice. Weakness does not attack where there is strength. The conviction of a single individual is the greatest strength the universe has to offer.

There lies the true temple of faith. It is in the shrine of goodness, the edifice built of kindness, the temple of patience and helpfulness and graciousness and compassion. It is to trust our fellow man, even when this is difficult and it is easier to fear him. It is to trust and to love one another even when the other is afraid of love. It is to not desert a friendship in the face of difficulty, to not abandon a hope because we are made to suffer. Faith is not to let others power over us, rather it is to hold the power of the soul by having the courage to love even when the other is undeserving. We do not negate an agreement, nor do we seek to power over those weaker than ourselves. But we do move our personal reality through faith in a way that reflects our greater being. In that is the true power of faith. The true temple of faith is the consciousness of our soul.

There may be great hardship in the test of this faith. We may be subjected to cruelty and fear. Yet, there lies the greatest reward. It is how we prove ourselves in terms of our identity in the face of peril, the "who am I" rising to meet the challenge of our being. It is how we prove we have the soul, that we have faith, that we can choose to be conscious human beings rather than fearful victims. It is the condition of our loving one another even in the face of terrible hardship. There is no greater magical rite than this. There is no temple more beautiful than one dedicated to a human being able to love with his or her being. It is in the temple of our human being where is most visible the soul.

To help and to give in the face of hardship is to test the soul. To tend the ill, to heal where there is fear, to soothe where there is misfortune, to bear bravely tragedy where others bear confusion, to love even in the face of rejection; these are the paths to the soul. To temper rage, to encourage kindness in others, to be charitable, to care; these are the paths that raises us from our more primitive past. It is how we rise above the forces of chaos with the forces of a universe becoming itself. To be free of angers, to be steadfst in the face of opposition, to protect the weak and to not cower in the face of coercion; these are the marks of a being whose level of consciousness has raised him or her into a new world of trust and agreement. How great is faith, a given word sealed with a handshake. These are in the power of the soul.

To be joyful with our reality and to suffer without it. These are the rituals demanded of us who are in the soul. It is ritual as old as time and as universal as the experience of all living things. How simple and yet how complex. If our ancestors built vast temples to the soul, to their gods, we need not. We need but have faith, to be still, to pray, to meditate, to believe. These are the expressions of our ability to choose when we are posed with hardship and suffering. It is how we personify our reality. In each one of us is "I am that I am." It is how we lend the soul to reality. If we must build shrines, then let them be built to this. Then we can touch our world with joy and the soul.

When we touch with the soul, our reality changes for us. When we dream from the soul we create new worlds. When we pray with the soul, we sow goodness and love to spread like a mantle throughout our world. When we are in the soul, we are guardians of a great energy of universal order that is manifest in our being as love. When we look into our soul and see our universe there, then we are looking at the world with eyes of love. It is to understand with the clear logic of faith, to believe with the courage of sanctity of being. To wish another well is to leave behind the essence of the power of the soul, to bring joy into that other's world. To pray for another is like a mist that will linger and cover with its mantle. It is the energy of the soul backed by faith.

If the ancients had tried to capture the essence of the soul, in effect to coerce it and bottle it and then use it in magical charms, they failed to capture its main essence. In their efforts to capture the soul, they were destroying that which they were trying to hold. The power of the soul is negated through trespass against it and can flourish only when it is released with love and belief. Coercion undoes the soul, whereas the free mind of man giving of itself of its own will is rich in the soul. This is important, for as we understand the soul better in the future, we must guard against repeating the errors of our ancestors. As we believe in the soul, it gains in consciousness and we will better manifest its power. But this cannot come about with coercion.

To speak the truth, to be thoughtful of others, to turn away from vanity and flattery, to be thankful and grateful to others, to avoid gossip and to not injure others; to genuinely wish another well, to love our fellow human beings as well as other living creatures, to celebrate life. These are the powerful forces of the energy of faith. We need to do these with selfessness, with sincerity, with openness. It is to step from the shadow of the soul into its brilliant light.

When we reach out and touch another, we are touching a soul. When we become conscious of this, we do it with reverence and awe, with passion and love. The power of the universal order flows where there is a complete surrender to this, to surrender to a being with a soul. "I am my being" is more powerful than we can know. Worlds meet. The light of the soul shines between them.

The soul's presence illuminates the darkness, for it gives strength and faith in the face of defenselessness and fear. In the face of evil and ill-will, it is faith that sustains us. The power of belief, of human identity, of an interrelated infinity tied into each one of our existences at birth, all define our being capable of the soul. When we become conscious of this, we illuminate the darkness that would do us ill and overcome it with the brilliance of our human love. It spreads like a mist of light into our world, and thus we pray from the very depth of our existence. Our presence personifies the world we live in. This is how free human beings overcome evil.

When we are truly free as human beings, free to love one another, free from trespass, free to do with joy as we will, then we are in the kingdom of the soul. Then the universe will be free to work with us and around us will materialize a reality of plenty. It will become a world of plenty, of goodness, of greater abundance and good-will than we had ever known. When we asked: 'Is there a natural order . . . ?" we have seen that we can look into our minds and see the universe there. We are the creators, the communicators, the energy of being that harnesses the power of the stars into our personal reality. We are the "mind" and we are "love." We are the focus of a universe made into a microcosm of flesh: we are the soul. When we look up at the sky and wonder: "Who am I?"; the answer that forms itself is: "We are Free!" Can we dare to believe in God?

As we gain this freedom, as we learn to do with love and gentleness, as we spread kindness and patience and faith; as we gain the soul, we will step into a future world of a conscious man, woman in a conscious world. What brilliance will await us? To be one with the world, to be as one with all living things, to be with the wind, to be powerful in the soul, to be with the energy of a world moving itself. If we look into another's eyes, feel the warmth of another's hand, stand in another's living presence; there is a whole universe manifest there! With childlike awe and simplicity, it is to see the soul in a flower petal, or the universe in a drop of rain. How wonderful! We live and we die. Yet, when we see this, we know we live. What great creature is this that is called man, woman? We dream, we love, we do; and thus we project ourselves into this world. It is the incredible dream of being born: "I Am!"

To have looked into a universe and having seen the world there; to have looked into a flower and seeing another's eyes there. To see being created of itself, from three. To see belief move reality. These are the power of Faith. To believe, to love, to choose, to be in the soul. These are how we gain the soul.

"I am Man."

Chapter Twenty-Nine

It Is Important to Love One Another

IT IS IMPORTANT to love one another, for that is how the soul of the universe is brought into our world. It is how that vast energy of All that Is is brought into the being of each one of us. It is how the soul's infinity is focused here in our lives. To love is to experience the mystery of the soul with our being. It is to be born, to personify our existence, and to move reality with our being. When we love one another, we join with the divine joy of a universe moving itself.

We have traveled far from the first probings of reality with our mind. From the material probings of three points in space to how our mind is a greater reality than that of its own consciousness; from the idea that society is a creation of mind to the idea of Habeas Mentem which shows that society is but an extension of the universal order when the mind is free. We have seen that, through agreement in a world free of coercion, the mind invites into itself its own individual definitions of how it is at infinity, its identity. Through exchange by agreement, we thus share in each other's identities. We have seen that, through interrelationship, the most complex is reduced to the simplest, for the image of infinity is ingrained in every part of itself. Thus, it is evident in all we do. We have traveled far from seeing that to be in the mind, when free, defined for us our personality in terms of how it occupies its own definition in time and space. And that to be conscious of this, in terms of Who we Are, is how we gain the soul. We have gone from a totally material image of the universe as an interrelated infinity to a totally spiritual universe as a definition of the human soul. We have spanned the material universe on the left side of the brain with a linear logic describing an interrelated infinity only to return on the right side of the brain describing the irrational logic of belief. Though we have travelled far in relation to our original three points in space asking the question: "Is there a natural order?", we have seen that to be in the mind in terms of the universal order, to be in agreement with oneself and one's fellow man, is the greatest achievement of the universal order. Now we can see that to have the soul and to love one another is the greatest power of the universe in our individual, daily lives. It is to personify reality in terms of our greater being in Who we Are.

On a social scale, the universe's order enters the world in terms of how we are as a society free from coercion, or in terms of how we exchange with one another through agreement. On a personal scale, the universal energy enters our personal being in terms of how we have gained the soul, or in terms of how we are conscious of our being and how we love one another. When we are conscious as beings in All that Is, reality expresses itself as love between human beings. This is the mystery of the soul: What is the image at Infinity that can define itself in man as Love?

If we look into another human being's eyes and see a fellowship there, if we love them, then we are looking at them with the soul. We are looking at them with our eyes as they are seen from the vast definition that is our personality, looking into their eyes as their personality is defined there. Be they dark and warm or blue and deep, they are the eyes of a fellow soul. Whether stranger or friend, brother or lover, we know them. In some distant and mysterious way, we know one another and there is recognition of this. We wish to meet again, to know one another, and we manifest our beings in friendship. Identities meet on a vaster scale. Souls meet. When souls love one another, universes meet.

If what meets the eye is but a microcosm of far greater images, how erotic to have bodies meet, to hold each other and press against each other's being the warmth of our flesh. A vast crucible of universal order brings together two beings, holds them in each other's thoughts until they join together in friendship or as lovers. What magic force lies behind the mystery of how we meet, of how we love one another? If in the here and now are but representations of greater identities, then we already know them, and their meeting is more than merely random events in our beings. What divine mystery lies behind the joy of people brought together by love?

To wish another well, to want them to have the best, to bring joy and laughter and pleasure into their lives, to spare them grief; these are gifts from the soul. They are the myriads of small sacrifices we make for the ones we love. It is to be willing to give without reward. It is to love even if there is risk of failure and rejection, selflessly. It is as if to come into an old familiar place, to risk being alone in one's thoughts, to have a love remain secret. Yet, it is to love with the hope that the other will love us too. How strong is a love that can risk loving another with the soul?

There is a greater image that would have us love one another thus. It is the image for which the soul reaches out, from which it had created us, and with which we are creating in our greater reality. It is perhaps the greatest expression of universal order, of the energy of coming together, of agreement, of mutual joy and caring for one another. It is the energy, as in belief, that would have us give of ourselves to another without expectation of return, to give freely because one needs to give. It cannot be forced, nor coerced from us. It is entirely a free expression of our soul. For this it is strong and cannot be weak, for it cannot be coerced from us. This giving is not self-seeking, for it is willing to forgo any reward; it is not slavish or small for it rejects self pity, even if its love is unrecognized. This love is greater than our little concerns for this world, for it is the love of a soul.

We gain the soul when we pass through this world and love in it. When we have reverence for all we do, when we are thankful and grateful for our daily being, when we seek what is beautiful and uplifting to us, when we dream, we are seeking with the soul. These are choices that are willed, they are pleasing to us because we seek to do good, to help rather than hurt, and to absorb misfortune rather than to pass it onto another. We are fearless in the face of coercion and are free in conscience to have faith. We become one with our existence,

positioned squarely within it at its center through the power
of our faith. There, we love one another and with that love bring
its power into our everyday being. By overcoming coercion and
by bringing the gentleness of love into our lives, we are inviting
that greater image into our world. In time, that which destroys
love, that which brings fear and disease and pain, becomes
removed. Thus, the great Joy of a universal order, of All that
Is, is brought into our world.

How magic! How sublime! To find love in another. It is not
a free good that comes from merely wishing for it. Rather, it
is a precious gift, beautiful and rare and valuable because of
its scarcity. It is not enough to command that we love one
another. This would not and should not work. We need to find
love, to discover it with that same childlike simplicity that defines
infinity. It must be gained, we work for it, it is elusive and exciting, for it is a precious gift from All that Is. It is enriched with
passion, perhaps even consumed by it as if by fire, yet it remains more powerful in its greater crucible of the soul. To
touch, to give, to have a loving thought, to desire; they are forces
of the energy that is our love. To reach out for another human
being, for their hand that is willing, is to reach out into their
private universe. And when they return this love, they reach
for us with their soul. It is in the love of fathers and mothers
for their children, of a family for each other. It is in the magical
love of lovers. It is in the caring love of friends. In each is the
divine image of a universe reaching for us with the soul.

To thrill at the mystery of a new born child, at the contact
of a new being brought into the world. We have all traveled
together and are alive on this small planet in this vast cosmos.
Each of us from our separate life is nevertheless part of the
sphere's greater being. We can laugh and play and amuse each
other and know each other though we are all different. Some
are tall, others shorter; some are light in coloring, others are
dark. Some are painfully brilliant, others are comfortably dull.
Yet, we are all human and in this should be revealed our goodwill to one another. To love one another should be shared on
a mutual trust, on a mutual rejection of fear and coercion of
one another, and thus based on a mutual friendship. When we
trust our lives in the energy of the soul, this is a possibility.

We trust in life when we are in the soul, for we then understand it in ways that are illogical to our mind, yet which we move with our greater being. In the end, it is we who create our reality. We are all on this planet together. There lies the image at infinity that would have us love one another.

With childlike simplicity, we can invoke the power of a universe, when we love another. When we love a tree, and have reverence for it before it is cut; when we love the Earth and its soil and mountains and rivers and deserts, and we have care for them before we change them in any way; when we touch the being of another and are very sensitive to their existence and feelings and consequences, and we wish them well . . . These are the times that we are loving them with the soul. We cannot reverse the presence of a tree that had stood for decades or centuries without being aware of its being. We cannot step into another's life without being keenly aware of our presence in their being. We cannot reach into the life forces of this planet without being aware of the possible disturbance this may do. So we approach with great caution and care. We must be careful when we pass through this world, for we risk damaging our soul. There is a power of the universe that is invoked when we do these. It becomes evident in what it is we leave behind.

Earth is naturally a beautiful planet. Though it had been a hard existence, difficut, and painful, we survived and beautiful souls had populated it before. They had eased its harshness into its present beauty. How will we leave it for our future? If millenia had traveled on the energy of the past, how will millenia look back when it is we who had passed through? What will the universe have materialized here from our being? This is what we must think of when we love one another, for these are the images we are inscribing in Infinity.

There lies the faith in the soul. Immortality is not given to us in this existence, but the soul lives on in our universal identity. How we pass through this world is how we will enrich or decay that identity with our deeds. To believe in the soul is to have the courage to let it work for us in its many mysterious ways. With good deeds, the soul will grow, and in this growth it will become more powerful in terms of our deeds. With faith in the soul, that greater power will in turn grow in how reality will influence us. Good deeds, good living, good beliefs, all strengthen the soul. It is to learn to think of our being not merely with the mind, but to think of ourselves with our total being. It is to move reality from this being, and thus to be more powerful in it because we love. When we are free from coercion, when we do not coerce others and are instead loved by them, then our world moves in mysterious ways that come about only through a faith in the soul. But this must be chosen and this faith must be willed. It happens naturally when we love one another. And thus we live on with what we leave behind.

It is important to personify our existence with love. How we love is how we are as a personality. It is how we dream, how we do, how we wish for others to do to us. But it is also how we eat our food, how we prepare the living things that must die for us, how we destroy; best if as little as possible. It is in how we care for those who are dependent on us by the virtue of their being weaker than we are. It is to rise in stature to not hurt any other thing, but rather to help whenever we can. These are the greater virtues of the soul that help us lend our personality to our reality and, thus, to the reality of others. But they are never invasive of others, for we love them only as we can without trespassing on their souls.

The joy of a universe thus becomes magnified in this love. If our existence in the here and now is but a microcosm of our greater being, then to love this greater being as we love one another as souls as well as personalities of human beings, is to cast our love off into the cosmos that is our greater universe. It is to worship in the soul there. To live our daily lives lovingly is to cast a magic force of universal love back out into infinity back down into our being. Through man the universe is creating Itself in its woman and man greater Image as the Joy of an Infinity looking back on Itself. How great is the Image of Love is All that Is as Love? We came from there.

So this is our Given Word: To Love One Another. It is to love with passion as well as compassion, to love from the soul as well as from the body. It is to love one another in all the ways that manifest for us the forces of a universe moving itself. The joy that will flow from this love is a world we may have yet never seen. It is to be strong, to be beautiful, to be truthful, to be trusting; all are the power of a universe defining itself through being. By being Who we Are in terms of how we are, we help demystify this love and bring it into our everyday reality. By bringing it into our reality through our individual being, we are working with a universe that is working to create itself through love. This is the magic power of how we choose to love one another. It is for this reason we must be free.

Will the ideas of Habeas Mentem and of the Soul overcome tyranny on Earth? Probably not. Tyranny is a manifestation of belief in the human soul as being other than it really is. Within these pages was written but a communication that would help us see ourselves as we really are. Then, through faith in our existence, in seeing ourselves as we are in terms of our reality, we will then work to overcome tyranny. Then, men and women will cease to oppress one another, for they will have faith in their world and in their beliefs. They then will no longer be afraid to love one another.

Will the ideas here written unite all societies and all peoples and all governments into one? Probably not. The great diversity of peoples and cultures and forms of rule are as varied as are the places of the planet. And so they should be, for the universe has enough diversity in it for all of us to be individually unique. Yet, there lies our unity, for in this freedom from all having to be the same, there arises a brotherhood of men and women who have the faith and courage to let each other be free. It would be the same with our religions and beliefs. All are manifestations of the soul. The only error we must guard against is the mistaken belief that it is acceptable for one man's belief to be forced on another. That is never so. It is the allegorical forbidden fruit of Eden. In its mistaken belief lies man's fall. We do have a soul, and it is because of this soul that we have the mind and the right to occupy our own space in time as we believe. When this is agreed upon, then we are free to have the soul. Then all societies and all peoples and all governments and all forms of beliefs will merge into one. But that will happen because it was willed and chosen by free individuals and not because it was imposed on them by another's belief. In that is the power of the soul.

Then, will there be no more war? When we love one another,

the planet will change again: Yes.

And what of the future? Will the society of the future be governed best by the rule which governs least? When we learn to work by agreement and no longer trespass on one another, yes. And of other worlds? Will they reach for us as we reach for them? When we have the soul, yes. In the Totality Image of All that Is, the world will change again.

Who is Man, Woman? We are how we gain the Soul, how we love one another.

"Imagine a pebble lying on the beach ..."

"Who am I? ..."

"Let us remember, as if in a dream ..."

We came from there.

* * * * * * * *

HAVE THE SOUL.

Feb. 12
The living waters, not the dead earth. It is as if the dormant earth opened its dark and living eye upon us.
                              [Henry David Thoreau's Journal]

Three things no man can alter: the stars in their courses, the flow of the tides, and the pattern that unrolls from the given word.
                              (The Pendragon. Catherine Christian)

Human freedom is the first postulate of practical reason . . .
The immortality of the soul is the second postulate of practical reason . . .
The existence of God is the third postulate of practical reason . . .
Who are we? . . . If God exists, then there is an answer.
                              Hans Kung, *Does God Exist?*